David Seal (277)

The theory of groups

IAN D. MACDONALD

The theory of groups

OXFORD · AT THE CLARENDON PRESS

Oxford University Press, Ely House, London W. 1
GLASGOW NEW YORK TORONTO MELBOURNE WELLINGTON
CAPE TOWN IBADAN NAIROBI DAR ES SALAAM LUSAKA ADDIS ABABA
DELHI BOMBAY CALCUTTA MADRAS KARACHI LAHORE DACCA
KUALA LUMPUR SINGAPORE HONG KONG TOKYO

ISBN 0 19 853138 9

First Published 1968
Reprinted from corrected sheets of the first edition 1970, 1975

Printed in Great Britain
at the University Press, Oxford
by Vivian Ridler
Printer to the University

Preface

THE present work sprang from a course of lectures given to an Australian pass class, when I found that no existing elementary text really met my needs. It bears many marks of its genesis in that fiery crucible. I refer first of all to the presentation of even the most elementary ideas in detail and with the support of many illustrations and examples. Next, I have in mind the single-minded concentration on groups, to the exclusion of links with other mathematics and of applications; experience shows that a course of this kind has to be self-contained as well as simple, and a glance at the Appendix will show how little material is assumed known: even Zorn's lemma and its cognates are passed over in silence and never used. The third point is the careful collection of problems for the reader to solve, some invitingly easy, others difficult or worse. Naturally my pass students reached their depth after only a few chapters. The rest of the book followed on logically, and the subject-matter in particular selected itself. No doubt the outcome is open to many criticisms, but I hope that they will be tempered by the experience of giving a course such as mine.

It seems to me evident, if not self-evident, that even the humblest university student should be exposed to a little abstract algebra. The problem of where to break into the subject has an attractive answer in the theory of groups. But this belief, and the consequent existence of this book, must not be construed as a disparagement of any other branch of modern algebra. I feel that, in particular, the theory of group representations is of the utmost importance in developing the theory of groups. This book contains as much group theory as I consider it advisable to study in isolation; after mastering it the enterprising reader is urged to enrich his knowledge by broadening his algebraic horizons. It may be well to note here that a modern honours course must contain much other abstract algebra of importance.

A few comments on the mathematics of the book are addressed to the expert. An honest attempt has been made to place equal emphasis on finite and on infinite groups. Minimal axioms for groups have been taken, both to give the novice exercise and because they have a place in a work of this type. The main text does not contain the new proofs of the structure theorem for finitely generated abelian groups and of the three Sylow theorems: my feeling is, on pedagogic grounds, that to the beginner they are hardly easier and probably less enlightening than the classical proofs. The problems are usually common currency among group theorists when they are not standard.

Finally, some words of gratitude. A number of my colleagues and students have read the manuscript in part and made suggestions concerning it. During the summer of 1967–8 the University of Queensland made available a Junior Research Assistantship, and Mr. J. R. Ellwood, who occupied this position, read the manuscript with care and solved all the problems. The typists were Mrs. Margaret F. Boden of the University of Newcastle, New South Wales and Miss Pauline C. Baker of the University of Queensland; their efforts were considerable, skilful, and always cheerful, and Mrs. Boden in particular produced many fine stencils. In the last stage of the work I have become much aware of the high professional standards and competence of the Clarendon Press. My expectations in this respect were more than fulfilled, and the officers and staff have been kind, helpful, and courteous without fail. To all these good people I offer my thanks.

I. D. M.

The University of Queensland
Australia
April 1968

1

Some examples of groups

SINCE examples are of great importance in the theory of groups we devote this chapter to a survey of the examples that are commonly encountered and used. We must of course indicate which of the several possible definitions of 'group' we have in mind, and the reason for our particular choice is simplicity in proving that certain systems we shall present are in fact groups. If the reader is familiar with most of the facts in this chapter he may yet find it of value as a systematic revision of his knowledge.

DEFINITION. A *group* is a set and a binary operation† in that set satisfying certain conditions;‡ that is, the elements a, b (in that order) in the group G define an element usually written as ab and usually called the product of a and b, the binary operation being (conventionally) called multiplication. The conditions are:

(i) G is closed under multiplication.

(ii) The associative law for multiplication holds, that is

$$a(bc) = (ab)c$$

for all elements a, b, c in G.

(iii) There is at least one identity element in G, that is an element e such that $ae = a$ for all a in G.

† For more details about binary operations see the Appendix.

‡ The reader who wishes to assume more, for instance that identities and inverses have some of the two-sided and uniqueness properties enumerated in Theorems 2.01–2.05, may do so. He will not then have to prove the corresponding theorems, but he will have the additional task of verifying that all the examples presented in this and subsequent chapters are groups in the sense of his definition.

(iv) Some identity element e has the following property: given any element a in G there is at least one solution x to the equation $ax = e$, each solution being called an inverse to a and written as a^{-1}.

It is important to note that we are not assuming uniqueness of identities and inverses in a given group. Another fact that requires careful consideration is that we are using what might be called right identities and right inverses. Suppose we make just one change in the above axioms, replacing the equation $ax = e$ in (iv) by $xa = e$, so that (in an obvious sense) we are dealing with right identities and left inverses; then many systems that are far from being groups satisfy the new axioms. We may for instance take any set S, such as the set of all teapots, and define products by putting $xy = x$ for each pair x, y of elements of S. Axioms (i) and (ii) are readily verified, and *every* element y of S serves as an identity as required in (iii). To solve the equation $xa = y$ for x with a given y, we simply take $x = y$. The axioms are therefore satisfied if S non-empty. But S does not constitute a group if it has more than one element, for the equation $ax = y$ has no solution x if $y \neq a$.

We pass on to examples of systems that are groups. The most familiar and readily available are groups of numbers.

Example 1.01. The set Z of all integers with addition as its operation forms a group. To hold fast to the conventions in our definition would involve writing $3+7$ as 3.7 and calling 10 the 'product' of 3 and 7. We therefore have every excuse to call the binary operation addition and to use the familiar notation in this special case. The group axioms are easily verified: (i) is simply that the sum of two integers is an integer, and (ii) is well known, plausible, and a consequence of any axiom system for integers. We can take the number 0 as an identity, for surely $a+0 = a$ for all elements a of Z. As for (iv), an inverse of a is $-a$ since we have, just as simply, $a+(-a) = 0$.

Example 1.02. The set Q of all rational numbers forms a group under addition, the proof being similar to that for Z.

Contents

Example 1.03. The set R of all real numbers forms a group under addition. Again the reader should be able to find a proof quite like that given for Z.

Example 1.04. The set C of all complex numbers forms a group under addition.

It may be well to mention some sets of numbers that do *not* form groups. The odd integers do not form a group under addition (though the even integers do); a sufficient reason for this is that (i) does not hold; another reason is that there is no identity element, which implies that (iii) and (iv) cannot hold. The non-negative integers do not form a group under addition for in their case (iv) does not hold (it is an empirical fact that many naturally occurring sets with multiplication are not groups because (iv) fails). The integers with the binary operation of subtraction do not form a group because the associative law is not valid; for instance $5-(4-1) = 2$ and $(5-4)-1 = 0$.

Since multiplication is the other familiar operation on (real and complex) numbers, it is reasonable to ask what sets of numbers form groups when the group multiplication is taken to be the ordinary multiplication of numbers. The set of all real numbers is not a candidate for group status in this sense, for it contains the number 0 and (since the number 1 is clearly an identity element) we require for axiom (iv) a solution x to the equation $0x = 1$; and because there is no such x we do not have a group.

Example 1.05. The set of positive rational numbers forms a group under (ordinary) multiplication. To see this, note that (i) and (ii) are easily checked, take $e = 1$ in (iii), and take $a^{-1} = 1/a$ in (iv); $1/a$ exists since $a \neq 0$. We remark that the non-zero rationals form a group for similar reasons.

Example 1.06. The set of positive (or of non-zero) real numbers forms a group under multiplication of numbers. As the reader should have by now acquired some facility in proving such a statement there is no need to give further details.

Example 1.07. The non-zero complex numbers form a group under the usual multiplication.

Example 1.08. The complex numbers of unit modulus form a group under the usual multiplication.

Example 1.09. For each prime number p there is a subset of the complex numbers that forms a multiplicative group of some interest. Let Z_p^∞ denote the following set:

$$\{z : z \in C \text{ and } z^{p^n} = 1 \text{ for some integer } n\}.$$

(Note that the positive integer n depends on the element z.) We verify the axioms individually for this important group. Let z_1, z_2 be elements of Z_p^∞ with $z_1^{p^n} = 1, z_2^{p^m} = 1$. Then $(z_1 z_2)^{p^r} = 1$ where r is the larger of m and n, so Z_p^∞ is closed under multiplication. The associative law holds simply because it is always valid for multiplication of complex numbers. Clearly Z_p^∞ contains the complex number 1, which will serve as an identity element. If $z^{p^n} = 1$ then $(1/z)^{p^n} = 1$, so if z is in Z_p^∞ so is $1/z$—note that $z \neq 0$—and $1/z$ will act as an inverse of z. Therefore Z_p^∞ is a multiplicative group.

A slight knowledge of complex numbers reveals that the elements of Z_p^∞ are precisely all those numbers of the form

$$\cos\frac{2k\pi}{p^n} + i\sin\frac{2k\pi}{p^n}$$

for $n = 0, 1, 2,...$, where k is an integer with $0 \leqslant k < p^n$.

Further important types of groups have residue classes instead of numbers as elements. A residue class modulo n is a certain equivalence class of integers;† thus the residue class $[a]$ modulo n is the set of all solutions x to the congruence

$$x \equiv a \text{ modulo } n,$$

where n is a fixed positive integer. More explicitly but less precisely (in the strict mathematical sense) we have

$$[a] = \{..., a-n, a, a+n, a+2n,...\}.$$

† As explained in the Appendix.

The set of residue classes modulo n is denoted by Z_n; it contains n elements, each being an infinite set of integers.

Example 1.10. We define a binary operation called *addition* on residue classes by the following equation:

$$[a+b] = [a]+[b].$$

This merely says that the sum of the residue classes containing the integers a and b respectively is to be that containing $a+b$. We ought first, however, to have verified that the result $[a+b]$ is the same *whatever* representatives we choose in $[a]$ and $[b]$, so that the definition depends only on the classes containing a and b, not on these numbers themselves. This we proceed to do. Take any element of $[a]$; it will be $a+kn$ for some integer k. Similarly an arbitrary element of $[b]$ will be $b+ln$. Thus

$$[a] = [a+kn], \quad [b] = [b+ln].$$

We require $$[a+kn]+[b+ln] = [a+b]$$

if we are to have a proper definition of sum. But

$$[a+kn]+[b+ln] = [(a+kn)+(b+ln)],$$

and since $$(a+kn)+(b+ln) \equiv a+b \text{ modulo } n$$

we have $$[(a+kn)+(b+ln)] = [a+b],$$

and so all is well.

The group axioms are easily verified now that the delicate matter of definition has been dealt with. Closure is obvious, the associative law is well-known or alternatively follows from the corresponding axiom for integers, $[0]$ is an identity element, $[-a]$ is an inverse for $[a]$. Therefore we have a group of residue classes, denoted by Z_n, with addition as its binary operation. Notice that it contains precisely n elements where n is an arbitrary positive integer. We therefore have a solution to the problem of finding a group with a given finite number of elements.

The subject of multiplication of residue classes and the formation of multiplicative groups requires some care. The usual definition of multiplication is the obvious one:

$$[ab] = [a][b].$$

It is essential to verify that this is a proper definition. This may be carried out in a way similar to that explained in the case of addition, the key step being the remark that

$$(a+kn)(b+ln) = ab+(la+kb+kln)n$$

and so
$$[(a+kn)(b+ln)] = [ab].$$

The reader ought to fill in all the details for himself.

Unfortunately this multiplication of residue classes will not generally yield groups. Just as the number 0 cannot occur in multiplicative groups of numbers so the residue class $[0]$ cannot occur in multiplicative groups of residue classes when $n > 1$, for how can we find an inverse element for $[0]$? Remember that $[0][x] = [0] \neq [1]$ for all x. There is another complication. Consider residue classes modulo 4, for which

$$[2][2] = [4] = [0];$$

and so the product of two non-zero classes may be $[0]$. We must exclude $[0]$ from any group, and having done that there are still cases when we do not have closure.

We can, however, salvage something.

Example 1.11. The set of non-zero residue classes modulo p forms a group under multiplication, p being any prime number. The associative law and the existence of an identity element $[1]$ are obvious facts, but axioms (i) and (iv) require care in their consideration. Let $[a]$ be a non-zero residue class modulo p. This is precisely the same as supposing that p does not divide the integer a. An elementary and well-known result of number theory† states that for suitable integers u, v we have the equality

$$au+pv = 1.$$

Thus in terms of residue classes modulo p we have

$$[a][u] = [au] = [au+pv] = [1].$$

† See the Appendix.

Now consider (i), the closure requirement, and suppose that $[a]$ and $[b]$ are non-zero residue classes for which $[a][b] = [0]$. It follows from the existence of u and from the commutative and associative laws that

$$[b] = [a][u][b] = [u][a][b] = [u][0] = [0],$$

and we arrive at a contradiction. Therefore (i) holds. Since we have found an inverse $[u]$ to $[a]$, we have also verified (iv). This shows that we have a multiplicative group whose elements are the $p-1$ non-zero residue classes modulo p.

All our examples so far have been groups with the property that every pair of elements commutes, that is ab is always equal to ba, and this, of course, has been a consequence of the commutative laws for addition and multiplication of real and complex numbers. We omit detailed verification.

DEFINITION. The elements a, b of the group G *commute* if $ab = ba$. The group G is *commutative* or *abelian* if every pair of its elements commutes.

A pair of elements of the form x, xx in any group commutes because of the associative law.

There are many groups that are non-abelian, and they are often best presented as groups of mappings of one sort or another. This seems to be their natural context. To make this remark precise, we recall that there is a natural multiplication of mappings and that this multiplication is associative; details are to be found in the Appendix. The name *semigroup* is given to any non-empty set with an associative multiplication under which it is closed. Thus every group is a semigroup with certain additional properties; note that axiom (iii) implies that no group is empty. On the other hand many semigroups (for example, the even integers with the usual binary operation of multiplication) fail to satisfy group axioms (iii) and (iv).

These remarks, together with the facts presented in the Appendix, should make the following theorem and its proof apparent.

THEOREM 1.12. *The set of all mappings of a fixed (non-empty) set into itself forms a semigroup.*□

To be quite explicit about the multiplication of mappings used here, let ϕ and ψ be two mappings of our fixed set X into itself; then $\phi\psi$ is defined by putting $x(\phi\psi) = (x\phi)\psi$ for each element x in X.

We now have a means of constructing many semigroups. For groups of mappings, however, we require inverses of mappings, and we recall that a mapping from a set X into X has an inverse if and only if it is both one–one and onto.† This explains our interest in the concept of 'permutation'.

DEFINITION. A *permutation* of a set is a one–one mapping of the set onto itself.

Example 1.13. The set S_X of all permutations of a (non-empty) set X forms a group. It is easy to see that S_X is closed, and (since S_X is a subset of the set of all mappings of X into X) associativity is a consequence of Theorem 1.12. An identity element of S_X is the permutation ι which leaves every member of X fixed, and which therefore has the property that $\rho\iota = \rho$ for each element ρ of S_X. Finally, since ρ is one–one and onto X, each permutation ρ has an inverse mapping ρ^{-1}, with $\rho\rho^{-1} = \iota$, and it is easily verified that ρ^{-1} is one–one and onto X, that is ρ^{-1} is a member of S_X.

This group S_X is known as the *symmetric group* on the set X. In a sense that will later be made precise, the structure of the group S_X depends only on the number of elements in X. Since there are $n!$ distinct permutations of a finite set of n elements, we see that S_X has $n!$ elements if X has n elements.

When X is finite and small it is practicable to write down the elements of S_X explicitly and to carry out exploratory calculations. If $X = \{1, 2, 3\}$ then S_X is

$$\{\iota, (23), (31), (12), (123), (132)\}$$

in the usual permutation notation. This particular S_X is non-

† See the Appendix.

abelian, since we have, using the usual multiplication of mappings,

$$(23)(31) = (132),$$
$$(31)(23) = (123),$$

so that $(23)(31) \neq (31)(23)$.

Example 1.14. Let R_X be the set of permutations ρ of the set X such that each ρ leaves all but a *finite* number of elements of X fixed. It turns out that R_X is an interesting group. To prove closure the reader should show that if σ_ρ is the set of elements of X that are displaced by the permutation ρ then $\sigma_{\rho_1 \rho_2} \subseteq \sigma_{\rho_1} \cup \sigma_{\rho_2}$. The other axioms are readily verified. This group R_X is known as the *restricted symmetric group* on the set X.

Example 1.15. Let A_X be the set of permutations ρ of X such that each ρ leaves all but a *finite* number of elements of X fixed, and has even parity (that is, ρ is the product of an even number of transpositions). Once again the reader is invited to prove the group properties of A_X under the appropriate multiplication. The group A_X is called the *alternating group* on the set X.

We next consider certain permutations of rather special sets, namely the sets of points making up Euclidean space of one, two, or three dimensions. Initially our account of these groups of mappings will be geometrical and tentative, and it should not be regarded as rigorous. After these intuitive considerations we shall examine groups of matrices, both those that arise when coordinates are introduced in Euclidean geometry and those that result from generalization.

Let $\mathscr{L}$ be a line—a one-dimensional Euclidean space. It is obvious in the intuitive sense that two sorts of congruence (that is, distance-preserving) mappings of $\mathscr{L}$ onto $\mathscr{L}$ exist:

(i) Translations—a mapping ϕ of $\mathscr{L}$ into $\mathscr{L}$ is a translation if ϕ moves every point a fixed distance to the right or left.

(ii) Reflections in a fixed point O—a mapping ϕ of $\mathscr{L}$ into $\mathscr{L}$ is a reflection if ϕ maps each point P into the point $P\phi$ which is on the other side of O but at the same distance from O.

Clearly there is a translation ι which acts as an identity mapping; clearly each translation has an inverse, which is again a translation; clearly each reflection ϕ has ϕ as an inverse, because $\phi\phi = \iota$.

Hence we find two more groups.

Example 1.16. The set of all translations of $\mathscr{L}$ (with the usual multiplication) forms an abelian group. It is not hard to visualize the group properties for this set.

Example 1.17. The set of all congruence mappings of the line $\mathscr{L}$ forms a group. A congruence mapping may be presented as a combination of suitable translations and reflections, which should enable the reader to see how the effect of a congruence mapping may be 'undone' and an inverse constructed. The other laws are easier to grasp. Among points that the reader might ponder on is the fact that the product of two reflections is a translation.

The congruence mappings of the plane $\mathscr{P}$ into itself are a little more complicated to describe. There are:

(i) Translations—a mapping ϕ of $\mathscr{P}$ into $\mathscr{P}$ is a translation if the segment joining P to $P\phi$ is of constant length and direction for all points P.

(ii) Rotations about a fixed point O—a mapping ϕ of $\mathscr{P}$ into $\mathscr{P}$ is a rotation if P and $P\phi$ are at the same distance from O and if the angle subtended at O by P and $P\phi$ is constant for all points P.

(iii) Reflections in a fixed line $\mathscr{L}$—a mapping ϕ of $\mathscr{P}$ into $\mathscr{P}$ is a reflection in $\mathscr{L}$ if each point P maps onto its mirror image in $\mathscr{L}$.

We note several of the numerous groups formed by sets of congruence mappings of $\mathscr{P}$.

Example 1.18. The set of all translations of $\mathscr{P}$ forms an abelian group.

Example 1.19. The set of all rotations of $\mathscr{P}$ about a fixed point O forms an abelian group.

Example 1.20. The set of all congruence mappings of $\mathscr{P}$ forms a group. Note that this group includes all rotations about every point O of $\mathscr{P}$.

In the case of congruence mappings of three-dimensional space $\mathscr{S}$ into $\mathscr{S}$ we indicate some familiar types:

(i) translations;
(ii) rotations about a fixed line;
(iii) reflections in a fixed plane.

It is a remarkable fact that any congruence mapping of $\mathscr{S}$ which is a combination of translations and rotations and which leaves one point fixed is equivalent to a rotation about a line through that fixed point.†

The following groups of congruence mappings of $\mathscr{S}$ are of interest.

Example 1.21. The set of all translations of $\mathscr{S}$ forms an abelian group.

Example 1.22. The set of all rotations of $\mathscr{S}$ about a fixed line forms an abelian group.

Example 1.23. The set of all rotations of $\mathscr{S}$ leaving a given point fixed forms a group.

Example 1.24. The set of all congruence mappings of $\mathscr{S}$ forms a group.

Many other important groups arise as sets of mappings which leave some geometrical figure invariant. We shall give one case which gives a group that we shall mention several times, referring the interested reader elsewhere for a fuller account.

Example 1.25. The set of all congruence mappings of $\mathscr{P}$ which leave fixed a regular n-sided polygon is a group. We mean that the polygon is unchanged as a whole, but not necessarily unchanged point by point; and of course n must be an integer greater than 2. If the centre of the polygon is O then the required

† A proof is indicated in Chapter 6 of *Algebra for Scientists and Engineers* by Hans Liebeck (Wiley, 1969). There is a careful examination of congruence mappings of $\mathscr{P}$ and $\mathscr{S}$.

mappings are rotations of $\mathscr{P}$ about O through the angle $2k\pi/n$ for $k = 0, 1, 2,..., n-1$, and reflections in the lines joining O to each vertex and to the mid-point of each side. It will be found that there are $2n$ such mappings and that these form a group, a so-called *dihedral group*.

We shall now try to introduce some mathematical rigour into our geometrical considerations. We take coordinates in the usual fashion, though we shall not try to relate all the groups that now force themselves on our attention to Examples 1.16–1.24.

Coordinates for $\mathscr{L}$ arise from the real numbers. Groups corresponding to the congruence groups 1.16 and 1.17 are:

Example 1.26. The set of all mappings ϕ_d, with d real, such that

$$x\phi_d = x+d$$

for all real numbers x, forms an abelian group. The verification is omitted.

Example 1.27. The set of all mappings of the previous example together with all mappings ψ_a of the type $x\psi_a = 2a-x$ forms a group. The verification is omitted.

In higher dimensions points are regarded as ordered sets of real numbers when coordinates are introduced, and we write such ordered sets as row vectors. Now an $n \times n$ matrix is essentially a mapping of the set of row vectors into itself, the row vector being multiplied by the matrix in the usual way. Let therefore $\mathscr{R}_n$ denote the set of all row vectors such as

$$x = (x_1, x_2,..., x_n),$$

where $x_1, x_2,..., x_n$ are real numbers; and let $\mathscr{M}_n$ be the set of all $n \times n$ matrices, so that an element A of $\mathscr{M}_n$ has the form

$$A = \begin{pmatrix} a_{11} & a_{12} & . & . & . & a_{1n} \\ a_{21} & a_{22} & . & . & . & a_{2n} \\ . & & & & & \\ . & & & & & \\ . & & & & & \\ a_{n1} & a_{n2} & . & . & . & a_{nn} \end{pmatrix}$$

with the entries a_{ij} real.

Then A defines a mapping of $\mathscr{R}_n$ into $\mathscr{R}_n$, namely the mapping which takes each x into xA where

$$xA = (a_{11}x_1+a_{21}x_2+\ldots+a_{n1}x_n, \ldots, a_{1n}x_1+a_{2n}x_2+\ldots+a_{nn}x_n).$$

Example 1.28. The set $\mathscr{M}_n$ of all non-singular $n\times n$ matrices with real entries forms a group under matrix multiplication. The closure law is obvious. The associative law follows, for example, from the facts that a matrix represents a mapping and that the product of two matrices represents the product of the associated mappings. The unit matrix serves as an identity element, and an inverse of the element A is the inverse matrix A^{-1}. So we have a group.

Example 1.29. The set of all real $n\times n$ matrices with determinant 1 forms a group under matrix multiplication. The least obvious point in the proof of this statement is the closure law.

Many variations on these examples are possible. In addition to having infinitely many possibilities for n, we can choose the rational numbers or the complex numbers or other sets of numbers for entries. We can even choose residue classes modulo some positive integer.

Example 1.30. The following set of matrices, in which i denotes a complex number for which $i^2 = -1$, forms a group under matrix multiplication:

$$\left\{\begin{pmatrix}1 & 0\\0 & 1\end{pmatrix}, \begin{pmatrix}-1 & 0\\0 & -1\end{pmatrix}, \begin{pmatrix}0 & 1\\-1 & 0\end{pmatrix}, \begin{pmatrix}0 & -1\\1 & 0\end{pmatrix}, \begin{pmatrix}0 & i\\i & 0\end{pmatrix}, \begin{pmatrix}0 & -i\\-i & 0\end{pmatrix}, \begin{pmatrix}-i & 0\\0 & i\end{pmatrix}, \begin{pmatrix}i & 0\\0 & -i\end{pmatrix}\right\}.$$

The verification is omitted. This group is called the *quaternion group*.

The multiplicative semigroup of all $n\times n$ matrices with (say) real entries contains many subsets which form multiplicative groups *not* containing the unit matrix. We exhibit one non-trivial specimen.

Example 1.31. The set of matrices

$$\left\{\begin{pmatrix} x & x \\ 0 & 0 \end{pmatrix} : x \text{ real and non-zero}\right\}$$

forms an abelian group under matrix multiplication. The proof depends on the fact that

$$\begin{pmatrix} x & x \\ 0 & 0 \end{pmatrix}\begin{pmatrix} y & y \\ 0 & 0 \end{pmatrix} = \begin{pmatrix} xy & xy \\ 0 & 0 \end{pmatrix},$$

which proves closure, shows that an identity element is the matrix in which $x = 1$, and indicates how to find inverses since

$$\begin{pmatrix} x & x \\ 0 & 0 \end{pmatrix}\begin{pmatrix} 1/x & 1/x \\ 0 & 0 \end{pmatrix} = \begin{pmatrix} 1 & 1 \\ 0 & 0 \end{pmatrix}.$$

Our final example is of lesser interest than multiplicative groups of matrices.

Example 1.32. The set of all $m \times n$ matrices forms an abelian group under matrix addition. The entries may be rational numbers, real numbers, complex numbers, etc.

Problems

(Harder problems are starred)

1. Discuss the following systems with a view to determining which of them are groups.

(i) The positive real numbers, with the binary operation of division.

(ii) The integers that are divisible by 10, with addition.

(iii) The non-zero rational numbers, with a^b taken as the 'product' of a and b.

(iv) $\{a+b\sqrt{2}: a \text{ and } b \text{ rational, not both zero}\}$ with the binary operation of number multiplication.

(v) The integers, with $2a+3b$ taken as the 'product' of a and b.

(vi) The rationals, with $a+b-ab$ taken as the 'product' of a and b.

(vii) The set of all 2×2 singular matrices with real entries, under matrix addition.

(viii) The set of all vectors in three-dimensional Euclidean space, with the binary operation of vector product.

(ix) The set $\{\phi_k: k \text{ real}\}$ of all mappings ϕ_k, where

$$x\phi_k = kx,$$

of $\mathscr{L}$ into $\mathscr{L}$.

(x) $\{a+b\sqrt{2}+c\sqrt{3}: a, b, c \text{ rational, not all zero}\}$ with the binary operation of number multiplication.

2. Which of the groups in the previous question are abelian? Which of the rest are semigroups?

3. Show that the non-zero residue classes modulo 6 do not form a group under multiplication.

Prove that if the non-zero residue classes modulo n form a group under multiplication and $n > 1$ then n is prime.

4. A set of matrices forms a group under matrix multiplication. Show that either every member of the set is singular or every member is non-singular.

5. Let p be any prime number. Show that the following set forms an additive group:

$$\{m/p^n \colon m \text{ and } n \text{ integers}\}.$$

Prove that the mapping ϕ for which $x\phi = px$, where x is an arbitrary element of the set, is a permutation.

6. Let G be a group of permutations on a set X. Show that those permutations which leave fixed a given element x of X form a group.

7. Let T_n be the set of $n \times n$ unitriangular matrices with real entries (that is matrices with 1 for each entry on the principal diagonal and 0 everywhere below this diagonal). Show that T_n is a multiplicative group.

8. Let T_n^* be the set of $n \times n$ unitriangular matrices with residue classes modulo p for entries (p is a fixed prime). Show that T_n^* is a multiplicative group, and find the number of elements in it.

*9. Let S be any set and G be any group. Prove that if multiplication of the mappings is *suitably* defined then the set of all mappings of S into G forms a group.

10. The set of all ordered triples of integers is denoted by G; thus a typical element of G is (α, β, γ) where α, β, γ are integers. Multiplication in G is defined as follows:

$$(\alpha, \beta, \gamma)(\xi, \eta, \zeta) = (\alpha + (-1)^\beta \xi, \beta + (-1)^\gamma \eta, (-1)^\xi \gamma + \zeta).$$

Show that G with this multiplication is a group.

*11. Show that there is a largest multiplicative group of 2×2 matrices with real entries containing $\begin{pmatrix} 0 & 0 \\ 5 & -5 \end{pmatrix}$, and find this group.

*12. A semigroup S has the property that for each element x in S there is an element denoted by x^{-1} in S such that $yxx^{-1} = y$ for all y in S. Is S a group?

**13. Let S be a semigroup with the following properties:

(i) There is an element e such that $ex = x$ for all x in S.

(ii) For each such element e and for each x in S there is an element x^{-1} for which $xx^{-1} = e$.

Prove that the set $\{se \colon s \in S\}$, where e is fixed, forms a group.

2

Basic theorems and concepts

THE theorems in this chapter are of two sorts. First, we must deduce some facts from the axioms. Second, we have to discuss the important concepts of homomorphism and isomorphism. The facts appear rather pedantic unless they are viewed against a background of systems that are not groups, while the concepts are of an importance that cannot be over-emphasized.

We recall our definition of a group as a semigroup with right identities and right inverses, in a sense made precise in Chapter 1. Sometimes a definition is given that states that each right identity is also a left identity and that each right inverse is also a left inverse; but we can deduce these facts from our axioms.

THEOREM 2.01. *If the element x of any group G has x^{-1} as a (right) inverse, so that $xx^{-1} = e$ for some (right) identity element e, then $x^{-1}x = e$.*

Proof. Let e be a (right) identity element of G of the kind mentioned in axiom (iv). We have to use the associative law in the form

$$x^{-1}(xx^{-1}) = (x^{-1}x)x^{-1}.$$

The left-hand side of this equation reduces to $x^{-1}e$ and so to x^{-1}, by axioms (iii) and (iv); therefore

$$x^{-1} = (x^{-1}x)x^{-1}.$$

But, by axiom (iv), x^{-1} itself has an inverse, denoted by $(x^{-1})^{-1}$, such that $x^{-1}(x^{-1})^{-1} = e$; and we have

$$x^{-1}(x^{-1})^{-1} = \{(x^{-1}x)x^{-1}\}(x^{-1})^{-1}.$$

The left-hand side is e by (iv); and the right-hand side is

$(x^{-1}x)\{x^{-1}(x^{-1})^{-1}\}$ by the associative law, and clearly this equals $x^{-1}x$. Thus we have proved that $x^{-1}x$ is the identity element e, as required. □

The above theorem implies that x and x^{-1} always commute. This is perhaps not surprising to us, for it happened in all the examples of the previous chapter. The reader should consider whether there is a corresponding theorem in a semigroup with right identities and left inverses, such as a set with xy defined to be x. (We discussed such objects in Chapter 1. Here is a specific instance. It will be found that if S is

$$\left\{\begin{pmatrix}1 & 0\\ a & 0\end{pmatrix} : a \text{ real}\right\}$$

then matrix multiplication in S is such that the product of two matrices is the left-hand factor.)

Another readily acceptable fact follows.

THEOREM 2.02. *If x is any element of any group G with identity element e, such that $gg^{-1} = e$ for all g in G, then $ex = x$.*

Proof. The previous theorem and the axioms are used as follows:

$$ex = (xx^{-1})x = x(x^{-1}x) = xe = x.$$

Thus the result is proved; x and e always commute. □

THEOREM 2.03. *If a and b are elements of any group G, and if there is an element x for which $xa = xb$ or an element y for which $ay = by$, then $a = b$.*

Proof. Suppose $xa = xb$. The existence of inverses gives

$$x^{-1}(xa) = x^{-1}(xb).$$

Use of the associative law and Theorem 2.01 gives

$$ea = eb.$$

Theorem 2.02 now gives $a = b$. The deduction from $ay = by$ needs only the axioms and being thus easier is omitted. □

This theorem gives us left and right cancellation laws, in an obvious sense. The reader should consider whether there are such laws in the set S of matrices mentioned above.

In all our examples, there was only one identity element and each element had only one inverse.

THEOREM 2.04. *In any group G there is only one identity element, and each element of G has only one inverse.*

Proof. Suppose we had two identity elements, say e_1 and e_2. Since e_2 is a right identity $e_1 e_2 = e_1$, and since (by Theorem 2.02) e_1 is a left identity $e_1 e_2 = e_2$. Therefore $e_1 = e_2$, and G cannot have more than one identity element.

Suppose that the element a of G has both x and y as inverses. Then $ax = ay$, and by left cancellation $x = y$; therefore a^{-1} is unique. $\square$

THEOREM 2.05. *In any group, $x = (x^{-1})^{-1}$.*

Proof. We have defined $(x^{-1})^{-1}$ as a solution y of the equation $x^{-1}y = e$; we know that each of e, x^{-1}, y is unique. But clearly $y = x$ is one solution to $x^{-1}y = e$, by Theorem 2.01. Since it is the only solution, we have $x = (x^{-1})^{-1}$. $\square$

The next result is of a different character. It is a generalization of the associative law that appeared as axiom (ii) and it states, roughly, that a product of group elements can be bracketed in any way one pleases.

THEOREM 2.06. *Let fixed elements $a_1, a_2, \ldots, a_n$ be taken in order in any group G. Then the value of the product of all these elements in the given order is unaffected by the sequence in which products are formed.*

Proof. The meaning of this theorem is subtle. For the purpose of illustration let $n = 3$. The statement then is that the elements $a_1(a_2 a_3)$ and $(a_1 a_2)a_3$ are the same, for there are only two possible ways of forming products; either form $a_2 a_3$ first or form $a_1 a_2$ first. The statement does *not* suggest that $(a_1 a_2)a_3 = (a_2 a_1)a_3$, which is certainly false in general. When $n = 4$ the possible products of a_1, a_2, a_3, a_4 in that order are $a_1\{a_2(a_3 a_4)\}$, $(a_1 a_2)(a_3 a_4)$, $a_1\{(a_2 a_3)a_4\}$, $\{a_1(a_2 a_3)\}a_4$, $\{(a_1 a_2)a_3\}a_4$, and no others. The theorem implies that these products are all equal.

We prove the theorem by induction on n. There is nothing to do when n is 1 or 2. Let $n > 2$. The inductive hypothesis is that products of fewer than n elements are well-defined, since the method of bracketing is then immaterial. We consider an element formed by taking products in the ordered set $\{a_1, a_2, \ldots, a_n\}$. We may write a typical product of $a_1, a_2, \ldots, a_n$ in order as

$$(a_1 \ldots a_r)(a_{r+1} \ldots a_n)$$

where $1 \leqslant r < n$.

Take two such products:

$$p_1 = (a_1 \ldots a_r)(a_{r+1} \ldots a_n),$$

$$p_2 = (a_1 \ldots a_s)(a_{s+1} \ldots a_n).$$

We may assume that $r < s$; for if $r = s$ then certainly $p_1 = p_2$. Thus we have $1 \leqslant r < s < n$ and we seek to prove that $p_1 = p_2$. Now

$$\begin{aligned} p_1 &= (a_1 \ldots a_r)(a_{r+1} \ldots a_n) \\ &= (a_1 \ldots a_r)\{(a_{r+1} \ldots a_s)(a_{s+1} \ldots a_n)\} \\ &= \{(a_1 \ldots a_r)(a_{r+1} \ldots a_s)\}(a_{s+1} \ldots a_n) \end{aligned}$$

by the associative law. Since

$$(a_1 \ldots a_r)(a_{r+1} \ldots a_s) = a_1 \ldots a_r a_{r+1} \ldots a_s,$$

the latter being well-defined by the inductive hypothesis, we have $p_1 = p_2$. □

The above result holds equally well for semigroups, as we have nowhere used group axioms (iii) and (iv). We shall use it continually in future without specific reference.

THEOREM 2.07. *In any group G the inverse of $a_1 \ldots a_n$ is $a_n^{-1} \ldots a_1^{-1}$.*

Proof. Clearly use of axioms (iii) and (iv) (that is 'cancellation') in

$$(a_1 \ldots a_n)(a_n^{-1} \ldots a_1^{-1})$$

gives e; note the tacit use of the previous theorem. If it is so desired, this statement can be refined in rigour by the use of induction, but this is unnecessary except in the strictest sense. Since inverses are unique by Theorem 2.04, $a_n^{-1} \ldots a_1^{-1}$ is the inverse of $a_1 \ldots a_n$. □

We now have to consider the question of indices. Already we know what a^{-1} means if a is any element of a group G. The general associative law allows us to give a reasonable meaning to a^n where n is a positive integer; we may formally define $a^1 = a$ and $a^n = a^{n-1}a$ for $n > 1$, for instance. Theorem 2.07, with $a_1 = \ldots = a_n = a$, yields in this notation that $(a^n)^{-1} = (a^{-1})^n$. If we now define a^{-n} to be $(a^n)^{-1}$, for any positive integer n, then we have $a^{-n} = (a^n)^{-1} = (a^{-1})^n$; let us also put $a^0 = e$.

THEOREM 2.08. *If a is any element of a group G and m, n are any integers then $a^m a^n = a^{m+n}$.*

Proof. This is an easy result when both m and n are non-negative. We give proof to cover all possibilities, starting with the special case $a^q a = a^{q+1}$. This is an easy consequence of definitions when $q \geqslant 0$. Suppose that $q < 0$, and put $q = -r$ for the sake of clarity. Then $a^{r-1}a = a^r$, and suitable manipulation gives

$$a^{r-1} = a^{-1}a^r,$$

$$a^{-r+1} = (a^{r-1})^{-1} = (a^{-1}a^r)^{-1} = (a^r)^{-1}a = a^{-r}a,$$

that is $a^{q+1} = a^q a$, as required.

Now consider $a^m a^n = a^{m+n}$ where n is non-negative. The case $n = 0$ is immediate, and we prove the cases with $n > 0$ by induction on n. Suppose that $a^m a^n = a^{m+n}$ is true for some $n \geqslant 0$. If we take q (above) to be $m+n$ then

$$a^{m+n+1} = a^{m+n}a.$$

By the induction hypothesis $a^{m+n}a = a^m a^n a$, and of course $a^n a = a^{n+1}$. So $a^{m+n+1} = a^m a^{n+1}$, and the inductive proof is complete.

Finally let $n < 0$, and put $n = -s$ so that $s > 0$. By the previous case $a^{m-s}a^s = a^m$, from which we deduce

$$a^{m-s} = a^m(a^s)^{-1} = a^m a^{-s},$$

and so $a^{m+n} = a^m a^n$. □

We note without proof another result that can be established by a similar consideration of special cases.

THEOREM 2.09. *If a is any element of a group G and if m, n are any integers then* $(a^m)^n = a^{mn}$. □

Our next task is to equip ourselves with concepts that enable us to start constructing a *theory* of groups. So far we have merely surveyed examples and drawn rather mechanical inferences from axioms. The idea of 'homomorphism' is our starting-point.

DEFINITION. A *homomorphism* of a group G into a group H is a mapping ϕ from G to H for which

$$(xy)\phi = (x\phi)(y\phi)$$

for all x, y in G.

Note that the product xy is formed by means of the multiplication in G, whereas $(x\phi)(y\phi)$ is formed in H. It is possible to illustrate mappings from G to H in diagrams of the following sort.

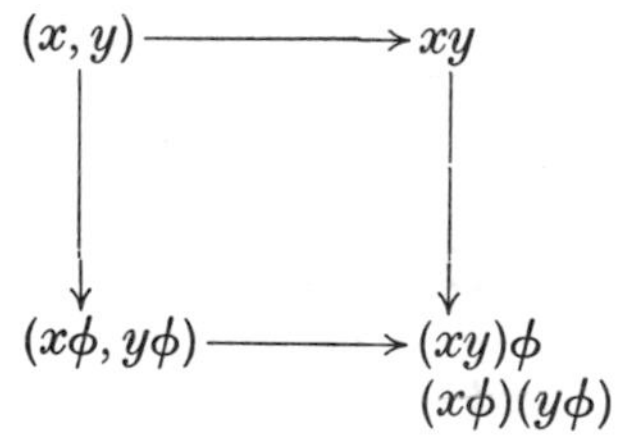

Here (x, y) is the ordered pair of elements x, y in G, and $(x\phi, y\phi)$ is similarly interpreted. The upper and lower arrows pointing right represent multiplication in G and in H respectively; thus multiplication in G maps (x, y) onto xy. The mapping ϕ clearly induces a mapping of $G \times G$ that takes (x, y) onto $(x\phi, y\phi)$; this is represented on the left. The right-hand arrow represents ϕ itself. Now ϕ is a homomorphism if and only if $(xy)\phi = (x\phi)(y\phi)$, that is if and only if the two paths from (x, y) to the opposite corner of the diagram have the same ending. This is what is intended by such statements as that 'ϕ is a homomorphism if it commutes with group multiplication'.

Example 2.10. Let Z and Z_n denote the additive groups of integers and of residue classes modulo n respectively. Define

a mapping ϕ from Z into Z_n by setting $a\phi = [a]$ for each a in Z. Since
$$(a+b)\phi = [a+b] = [a]+[b] = a\phi+b\phi$$
by the way in which addition of residue classes was defined, ϕ is a homomorphism.

Example 2.11. Let $\mathscr{M}_2$ be the multiplicative group of non-singular 2×2 matrices with real entries. We define a mapping ϕ from $\mathscr{M}_2$ to the multiplicative group R^* of non-zero real numbers by setting $A\phi = |A|$; here A is an element of $\mathscr{M}_2$ and $|A|$ is its determinant. It follows from the well-known fact
$$|AB| = |A||B|$$
that ϕ is a homomorphism, for
$$(AB)\phi = |AB| = |A||B| = (A\phi)(B\phi).$$

THEOREM 2.12. *The product of two homomorphisms is a homomorphism.*

Proof. This means that if ϕ is a homomorphism from G into H, and θ is a homomorphism from H into K, where G, H, K are of course groups, then the mapping $\phi\theta$ from G into K is a homomorphism. Let x, y be arbitrary elements of G. Then
$$(xy)(\phi\theta) = \{(xy)\phi\}\theta$$
by the definition of $\phi\theta$;
$$(xy)\phi = (x\phi)(y\phi)$$
since ϕ is a homomorphism from G into H; and
$$\{(x\phi)(y\phi)\}\theta = \{(x\phi)\theta\}\{(y\phi)\theta\}$$
since θ is a homomorphism from H into K. Hence we have
$$(xy)(\phi\theta) = \{(x\phi)\theta\}\{(y\phi)\theta\} = \{x(\phi\theta)\}\{y(\phi\theta)\},$$
which means that $\phi\theta$ is a homomorphism as required. □

THEOREM 2.13. *Let ϕ be a homomorphism from the group G into the group H. If e denotes the identity element in G then $e\phi$ is the identity element in H, and if x is an arbitrary element of G then $x^{-1}\phi$ is the inverse of $x\phi$ in H.*

Proof. Note that $x = e$ is the only solution to the equation $x^2 = x$ in G. Since $e^2 = e$, we have $(e\phi)^2 = e\phi$. It follows that $e\phi$ is the identity element in H. Similarly $xx^{-1} = e$ in G implies $(x\phi)(x^{-1}\phi) = e\phi$ in H. Since inverses are unique in H we now have $x^{-1}\phi = (x\phi)^{-1}$. □

Thus, in a sense, a homomorphism 'preserves identities and inverses'. Certainly some (but not all) of the group structure is preserved; our examples show that a homomorphic image of an infinite group may be finite, and a homomorphic image of a non-abelian group may be abelian. Even a homomorphism from one group *onto* another may involve some loss; see Examples 2.10 and 2.11. Hence the need for a stronger concept.

DEFINITION. An *isomorphism* from a group G to a group H is a homomorphism ϕ from G to H for which there exists an inverse homomorphism ψ from H to G.

These homomorphisms are inverse as mappings, that is, $\phi\psi$ is the identity mapping on G and $\psi\phi$ is the identity mapping on H. The groups G and H are said to be *isomorphic*, and we use the notation

$$G \cong H$$

to mean 'G is isomorphic to H'.

Of course the theorems on homomorphisms above apply to the special case of isomorphisms. Before further elucidating the nature of isomorphism, we study some examples.

Example 2.14. Let G be the multiplicative group of complex numbers

$$\{z_k : 0 \leqslant k \leqslant n-1\},$$

where $z_k = \cos(2k\pi/n) + i\sin(2k\pi/n)$ and n is an integer greater than 1, k being an arbitrary integer. (It can be verified that G is a group.) Let H be Z_n (Example 1.10). Define ϕ, a mapping from G to H, by setting $z_k\phi = [k]$. Since $z_k z_l = z_{k+l}$ by de Moivre's theorem, we have

$$(z_k z_l)\phi = (z_{k+l})\phi = [k+l] = [k]+[l],$$

and so ϕ is a homomorphism; in this calculation $k+l$ may have to be reduced modulo n but this makes no difference to z_{k+l}. Next we define ψ, a mapping from H to G, by putting $[k]\psi = z_k$;

by the previous statement this is a good definition. It is easily shown that ψ is a homomorphism. It is plain that $\phi\psi$ and $\psi\phi$ are both identity mappings, and so ϕ is an isomorphism.

Example 2.15. Let G be the multiplicative group of non-zero complex numbers and let H be the multiplicative group of matrices of the form $\begin{pmatrix} a & b \\ -b & a \end{pmatrix}$, where a, b are real numbers not both zero; it should be verified that H is in fact a group. Define ϕ by

$$(a+ib)\phi = \begin{pmatrix} a & b \\ -b & a \end{pmatrix},$$

where a, b are of course real. Then

$$(a+ib)(c+id) = (ac-bd)+i(ad+bc),$$

$$\{(a+ib)(c+id)\}\phi = \begin{pmatrix} ac-bd & ad+bc \\ -ad-bc & ac-bd \end{pmatrix},$$

$$\{(a+ib)\phi\}\{(c+id)\phi\} = \begin{pmatrix} a & b \\ -b & a \end{pmatrix}\begin{pmatrix} c & d \\ -d & c \end{pmatrix} = \begin{pmatrix} ac-bd & ad+bc \\ -bc-ad & -bd+ac \end{pmatrix}.$$

Hence $\{(a+ib)(c+id)\}\phi = \{(a+ib)\phi\}\{(c+id)\phi\}$, and so ϕ is a homomorphism. If ψ maps $\begin{pmatrix} a & b \\ -b & a \end{pmatrix}$ onto $a+ib$ then one may similarly show that ψ is a homomorphism too, and since $\phi\psi$ and $\psi\phi$ are identity mappings it follows that ϕ is an isomorphism.

The following result will be of much use, and could indeed have been used in the study of the examples above.

THEOREM 2.16. *The mapping ϕ from the group G to the group H is an isomorphism if and only if it is a homomorphism which is one–one and onto.*

Proof. If ϕ is an isomorphism then ϕ is a homomorphism which has an inverse. By a general result on mappings,† a given mapping has an inverse if and only if it is one–one and onto. Therefore ϕ is one–one and onto.

Suppose conversely that ϕ is a one–one homomorphism from G onto H. Then ϕ, as a mapping, has an inverse ψ, by the general

† See the Appendix.

result quoted above. If ψ is a homomorphism, then ϕ is an isomorphism, so we wish to prove that $(uv)\psi = (u\psi)(v\psi)$ for any elements u, v in H. There are unique elements x, y in G for which

$$u = x\phi, \qquad v = y\phi$$

since ϕ is one–one and onto. Then

$$(uv)\psi = \{(x\phi)(y\phi)\}\psi = \{(xy)\phi\}\psi$$

since ϕ is a homomorphism; next

$$\{(xy)\phi\}\psi = (xy)(\phi\psi) = xy$$

since $\phi\psi$ is the identity mapping on G. However,

$$x = u\psi, \qquad y = v\psi,$$

and so we have $(uv)\psi = (u\psi)(v\psi)$. Therefore ψ is a homomorphism, and ϕ is an isomorphism, as required. □

At this stage there is an easy proof that neither of the homomorphisms in Examples 2.10 and 2.11 is an isomorphism. It is sufficient to show that neither is one–one. In the case of Z and Z_n we have that $(kn)\phi = [0]$ for *all* integers k, and so ϕ is certainly not one–one. In the case of $\mathscr{M}_2$ and R^* we have $A\phi = (-A)\phi$, a statement which is adequate to show that ϕ is again not one–one.

THEOREM 2.17. *Isomorphism between groups is an equivalence relation.*

Proof. We refer to the Appendix for a discussion of equivalence relations in general. It is necessary here to demonstrate reflexive, symmetric, and transitive properties for isomorphism.

(i) Clearly the group G is isomorphic to G; the required homomorphisms ϕ and ψ may be taken as identity mappings from G to G.

(ii) Suppose ϕ is an isomorphism from G to H and let ψ be its inverse homomorphism, then ψ is an isomorphism from H to G (note that ϕ is one–one and onto). If we interchange G and H, and interchange ϕ and ψ, we find that H is isomorphic to G.

(iii) Suppose that ϕ_1 is an isomorphism from G to H and ϕ_2 is an isomorphism from H to K. Then by Theorem 2.12 $\phi_1\phi_2$ is a homomorphism from G to K. Further, $\phi_1\phi_2$ is one–one and onto since both ϕ_1 and ϕ_2 have these properties. Therefore $\phi_1\phi_2$ is an isomorphism, and G is isomorphic to K.

Because the relation of isomorphism is reflexive, symmetric, and transitive, it is an equivalence relation. □

The objects of study in a theory of groups are the equivalence classes of isomorphic groups, and the relations between these classes. Such a class, being all those groups isomorphic to a given group, is called an *abstract group*. Since the isomorphism relation preserves structure in a certain strong sense, though it destroys individuality, an abstract group embodies the structure common to many groups, which may, of course, be quite different in the nature of their elements. For instance, we have studied (Example 2.15) a group of complex numbers which is isomorphic to a group of matrices.

Now many of the examples in Chapter 1 were groups of mappings. Not every group *is* a group of mappings—a complex number is not a mapping, for instance—but it is natural to wonder if every group is *isomorphic* to a group of mappings. The additive group R of reals is isomorphic, in a very natural way, to the group of mappings described in Example 1.26; the correspondence that maps the real number d onto the mapping ϕ_d is an isomorphism, as one may verify. This is a special case of the next result, which is due to Cayley.

THEOREM 2.18. *Every group is isomorphic to a group of permutations of a suitable set.*

Proof. The group G is given, and we have to produce a suitable set. We take the set of elements in G. Next we have to define permutations of this set, one corresponding to each element of G. Let a be a fixed element of G, and define the mapping ρ_a of G into G by setting

$$x\rho_a = xa$$

for all x in G.

We show that ρ_a is a permutation. Firstly, ρ_a is onto G, for given b in G we find that ρ_a maps ba^{-1} onto b. Secondly, ρ_a is one–one; for if $x\rho_a = y\rho_a$ then $xa = ya$ and so $x = y$.

Now we prove that $P = \{\rho_a : a \in G\}$, with the usual multiplication of mappings, is a group.

$$x(\rho_a \rho_b) = (x\rho_a)\rho_b = (xa)\rho_b = (xa)b,$$
$$x\rho_{ab} = x(ab);$$

hence the associative law in G gives $\rho_a \rho_b = \rho_{ab}$ for all a, b in G. It follows that P is a group with identity element ρ_e and with $\rho_a^{-1} = \rho_{a^{-1}}$ for all ρ_a in P.

Lastly we show that G is isomorphic to P. Let ϕ be the mapping from G to P such that $a\phi = \rho_a$ for all a in G. Then ϕ is a homomorphism since

$$(ab)\phi = \rho_{ab} = \rho_a \rho_b = (a\phi)(b\phi);$$

ϕ is one–one since if $\rho_a = \rho_b$ then $e\rho_a = e\rho_b$, and so $a = b$; and clearly ϕ is onto P. Hence ϕ is the required isomorphism.

Therefore G is isomorphic to a group of permutations of a suitable set. □

The permutation group P, which has just been put in correspondence with the group G, is called the *right regular representation* of G.

We mention without proof that a similar theorem holds for semigroups—every semigroup with an identity element is isomorphic to a semigroup of mappings. (Of course, isomorphism of semigroups has to be defined formally to make this statement meaningful.)

Theorem 2.18 shows that a finite group (that is, a group with a finite number of elements) is isomorphic to a group of permutations of a finite set. It is an interesting fact that such a group of permutations is isomorphic to a multiplicative group of matrices. A minor result is needed in preparation for the proof of this theorem.

LEMMA 2.19. *If the sets X and Y are in one–one correspondence and G is a group of permutations of X then G is isomorphic to a group of permutations of Y.*

Proof. It is intuitively clear that the elements of X may be replaced by the corresponding elements of Y. It is left to the reader to supply a formal proof. □

It follows in particular that if X is a finite set of n elements and if G is a particular group of permutations, then G, regarded as an abstract group, is independent of the nature of the elements of X. It is convenient and usual to take $X = \{1, 2, \ldots, n\}$, and to denote the groups S_X, A_X of Examples 1.13, 1.15 by S_n, A_n respectively in this case. We call S_n the *symmetric group of degree n*, and A_n the *alternating group of degree n*.

THEOREM 2.20. *Every group of permutations on a finite set of n elements is isomorphic to a group of $n \times n$ non-singular matrices.*

Proof. The lemma allows us to choose the finite set of n elements. For our purpose we choose the set V of row vectors, or $1 \times n$ matrices, which have a single entry 1 and the other entries 0; thus V has n elements. Let these elements be $v_1, \ldots, v_n$, where v_j has 1 in the jth column.

Our given group G of permutations is isomorphic to a group of permutations of V, by Lemma 2.19. Let ρ be an arbitrary element of G; then we may and shall denote the image of v_j under ρ as $v_{j\rho}$ for $j = 1, \ldots, n$; here we are thinking of ρ as acting on V and equally as acting on $\{1, \ldots, n\}$. We now define $A(\rho)$ to be the matrix whose jth row is $v_{j\rho}$. Thus $A(\rho)$ is an $n \times n$ matrix for which

$$v_j A(\rho) = v_{j\rho}$$

for $j = 1, \ldots, n$. Clearly $A(\rho)$ is non-singular because its determinant is ± 1.

Next we prove that the set $\{A(\rho) : \rho \in G\}$ of matrices forms a group A under matrix multiplication. For each j we have

$$v_j\{A(\rho)A(\sigma)\} = \{v_j A(\rho)\}A(\sigma) = v_{j\rho} A(\sigma) = v_{(j\rho)\sigma},$$

σ being an element of G; but further

$$v_j A(\rho\sigma) = v_{j(\rho\sigma)},$$

and since $(j\rho)\sigma = j(\rho\sigma)$ for each j we have $A(\rho)A(\sigma) = A(\rho\sigma)$.

So we have a group of matrices with identity element $A(\iota)$ and with $A(\rho^{-1}) = A(\rho)^{-1}$, where ι is the identity permutation.

Finally we prove that the mapping which takes ρ in G to $A(\rho)$ in A is an isomorphism. We have already shown that it is a homomorphism. It is one–one. For suppose that $A(\rho) = A(\sigma)$, then by looking at the defined structure of $A(\rho)$ we find $v_{j\rho} = v_{j\sigma}$ for each j and it follows that $\rho = \sigma$. It is also onto A, by definition of A.

It follows that G is isomorphic to A, as required. □

The matrices involved in the proof had an entry 1 occurring just once in each row and in each column, with 0 elsewhere. Such matrices are called *permutation matrices.*

COROLLARY 2.21. *Every finite group is isomorphic to a suitable group of $n \times n$ non-singular matrices.*

Proof. This follows at once from Theorems 2.18, 2.20, which ensure that a finite group is isomorphic to a group of permutations and that the latter is in turn isomorphic to a group of matrices. □

Example 2.22. As an illustration of these results we produce a matrix group isomorphic to S_3. We regard S_3 as the set of all permutations of V where $V = \{v_1, v_2, v_3\}$ and

$$v_1 = (1,0,0), \qquad v_2 = (0,1,0), \qquad v_3 = (0,0,1).$$

Now apply the method of the previous theorem. The permutation (12) of $\{1, 2, 3\}$ corresponds to

$$\begin{pmatrix} v_2 \\ v_1 \\ v_3 \end{pmatrix} = \begin{pmatrix} 0 & 1 & 0 \\ 1 & 0 & 0 \\ 0 & 0 & 1 \end{pmatrix}.$$

Similarly the permutations (23) and (31) correspond respectively to

$$\begin{pmatrix} v_1 \\ v_3 \\ v_2 \end{pmatrix} = \begin{pmatrix} 1 & 0 & 0 \\ 0 & 0 & 1 \\ 0 & 1 & 0 \end{pmatrix},$$

$$\begin{pmatrix} v_3 \\ v_2 \\ v_1 \end{pmatrix} = \begin{pmatrix} 0 & 0 & 1 \\ 0 & 1 & 0 \\ 1 & 0 & 0 \end{pmatrix}.$$

The permutations (123) and (321) correspond respectively to

$$\begin{pmatrix} v_2 \\ v_3 \\ v_1 \end{pmatrix} = \begin{pmatrix} 0 & 1 & 0 \\ 0 & 0 & 1 \\ 1 & 0 & 0 \end{pmatrix},$$

$$\begin{pmatrix} v_3 \\ v_1 \\ v_2 \end{pmatrix} = \begin{pmatrix} 0 & 0 & 1 \\ 1 & 0 & 0 \\ 0 & 1 & 0 \end{pmatrix}.$$

Finally the identity permutation ι corresponds to

$$\begin{pmatrix} v_1 \\ v_2 \\ v_3 \end{pmatrix} = \begin{pmatrix} 1 & 0 & 0 \\ 0 & 1 & 0 \\ 0 & 0 & 1 \end{pmatrix}.$$

We have reached the most appropriate point at which to introduce another fundamental concept concerning mappings of groups, namely that of 'automorphism'.

DEFINITION. An *automorphism* of the group G is an isomorphism of G onto G.

To the uninitiated it may seem paradoxical that a group G could possess any automorphism other than the trivial one which is the identity mapping of G onto itself. We therefore illustrate.

Example 2.23. Let G be any abelian group. Define a mapping ϕ of G into G by putting $x\phi = x^{-1}$, for all x in G. The abelian nature of G ensures that ϕ is a homomorphism:

$$(xy)\phi = (xy)^{-1} = y^{-1}x^{-1} = x^{-1}y^{-1} = (x\phi)(y\phi)$$

by Theorem 2.07. Next we observe that $x\phi^2 = x$, for all x in G, by Theorem 2.05; this means that ϕ itself is an inverse homomorphism to ϕ. It follows that ϕ is an isomorphism, and therefore an automorphism of G.

Example 2.24. For a more subtle example, we take for G the symmetric group on some set X; for the sake of definiteness and simplicity let $X = \{1, 2, 3\}$ and $G = S_3$. Choose one permutation ρ of the set X; for instance, $\rho = (12)$. A mapping ϕ of G into G

is defined as follows. If x is an element of G, express it as $\begin{pmatrix}1 & 2 & 3\\ a & b & c\end{pmatrix}$ where $\{a, b, c\} = \{1, 2, 3\}$; then apply ρ to all the symbols in $\begin{pmatrix}1 & 2 & 3\\ a & b & c\end{pmatrix}$, obtaining

$$x\phi = \begin{pmatrix}1\rho & 2\rho & 3\rho\\ a\rho & b\rho & c\rho\end{pmatrix}.$$

The reader should verify

(i) that $x\phi \in G$;
(ii) that ϕ is a homomorphism; and
(iii) that ϕ has an inverse homomorphism.

We omit these details, merely asserting that they prove ϕ to be an automorphism of G. This example might repay some study and reflection.

Further examples will appear in context later. At present we prove an important property of sets of automorphisms:

THEOREM 2.25. *The set of all automorphisms of a group forms a group.*

Proof. Let G be a group and let $A(G)$ be the set of all automorphisms of G. Multiplication in $A(G)$ is assumed to be the natural binary operation for mappings; if $\alpha, \beta \in A(G)$ then

$$x(\alpha\beta) = (x\alpha)\beta$$

for all x in G. Clearly $\alpha\beta$ is an automorphism of G, because (as proved in Theorem 2.17) the product of two isomorphisms is an isomorphism. The associative law holds, for we are dealing with products of mappings. The identity element is that which leaves every element of G fixed, and the inverse of α is the mapping α^{-1}, which is meaningful because α is an isomorphism. □

We call $A(G)$ the (*full*) *group of automorphisms* of G.

Problems

1. Show that if k and n are positive integers then there is a homomorphism from Z_{kn} onto Z_n. For what values of k is this an isomorphism?

2. Prove that the symmetric group S_n on n symbols ($n > 2$) is homomorphic to a group with 2 elements.

3. The set Hom(G, A) is defined to be the set of all homomorphisms from the group G into the abelian group A. Show that if multiplication of homomorphisms is *suitably* defined then Hom(G, A) is an abelian group.

4. Prove that the set $\{[1], [3], [5], [7]\}$ of residue classes modulo 8 forms a multiplicative abelian group.

Find the number of elements in Hom(G, A) when G is the symmetric group S_3 of degree 3 and A is the above group of residue classes.

5. Let $\mathscr{M}_2$ be the multiplicative group of all non-singular 2×2 matrices with real entries. Let T be the group of all mappings ϕ_i of the form

$$x\phi_i = \frac{a_i x + b_i}{c_i x + d_i},$$

where a_i, b_i, c_i, d_i are real numbers for which $a_i d_i \neq b_i c_i$; multiplication in T is defined by

$$x(\phi_i \phi_j) = (x\phi_j)\phi_i$$

for all x. Show that there is a homomorphism from $\mathscr{M}_2$ to T. Is it an isomorphism?

6. Let G be the multiplicative group $\{1, -1, i, -i\}$ of complex numbers, and let H be the multiplicative group $\{[1], [3], [5], [7]\}$ of residue classes modulo 8. Show that G is *not* isomorphic to H although they are both abelian and have the same number of elements.

7. Show that the group of congruence mappings of $\mathscr{P}$ into itself which leaves fixed an equilateral triangle (see Example 1.25) is isomorphic to the symmetric group on the vertices of the triangle.

*8. Prove that the set of all non-singular 2×2 matrices with residue classes modulo 2 for entries forms a group isomorphic to the symmetric group of degree 3.

9. Show that the *additive* group C of complex numbers is isomorphic to the group formed by all matrices of the type $\begin{pmatrix} a & b \\ -b & a \end{pmatrix}$ with the operation of addition.

10. Let S be the set $\{x : x \text{ real}, 0 \leqslant x < 2\pi\}$, and define $x \oplus y$ to be $x + y$ if $0 \leqslant x + y < 2\pi$, and $x + y - 2\pi$ if $2\pi \leqslant x + y < 4\pi$. Prove that an abelian group results, and that this group is isomorphic to the multiplicative group of complex numbers of unit modulus.

11. Find a group of matrices isomorphic to the alternating group A_4 on 4 symbols.

*12. Find the full group of automorphisms of the symmetric group S_3.

13. A one-one mapping ϕ of the group G onto the group G is an *anti-automorphism* if $(xy)\phi = (y\phi)(x\phi)$ for all x, y in G. Prove that ϕ is an anti-automorphism of G if and only if $\phi = \alpha\psi$, where $x\alpha = x^{-1}$ for all x in G, and ψ is an automorphism of G.

*14. Is the multiplicative group of positive rationals isomorphic to the additive group of rationals?

*15. Show that if A is any abelian group then the mapping ϕ_n taking each element of A onto its nth power, for a fixed integer n, is a homomorphism.

Discuss whether ϕ_n is an automorphism or not in the following cases:

(i) $A = Z$;
(ii) $A = Z_p^\infty$ and $n = p$;
(iii) $A = Q$.

***16. Show that in any semigroup S the following conditions are equivalent.

(*a*) There is an element e such that $xe = x$ for all x in S, and for each such element e and for each x in S there is an element x^{-1} such that $x^{-1}x = e$.

(*b*) For each x in S there is an element x^{-1} such that $yxx^{-1} = y$ for all y in S.

3

Complexes and subgroups

THE purpose of this chapter is to consider subsets of elements, in a fixed group, and relations among subsets. As we shall deal mainly with abstract groups from now on, it is desirable to make a small change in notation; we shall use the symbol 1 uniformly for the identity element (or unit element, as it may sometimes be called) of any group in which the binary operation is named multiplication.

We have already met situations in which one group is contained, *as a set*, in another. For instance, the integers are a subset of the additive group of rationals. But not every subset of the rationals forms a group with the original multiplication. For instance, the odd integers are a subset of the additive group of rationals but do not form an additive group in themselves.

DEFINITION. A *complex* of a group is any subset of elements of the group. If C is a complex of G we write $C \subseteq G$, in the usual set notation.

DEFINITION. A *subgroup* of a group is a complex which itself forms a group under the original group multiplication. If S is a subgroup of G we write $S \leqslant G$.

There are two obvious subgroups of the group G. There is G itself, and there is the subgroup (it clearly *is* a subgroup) which contains only the identity element 1 of G and which will also be written as 1. Any other subgroup of G is called *proper*. We shall write $S < G$ to mean that S is a subgroup of G and $S \neq G$. The subgroup 1 of G will be called trivial and indeed the group with 1 element will itself be called trivial. Thus $S < G$ means that either S is a proper subgroup of G or S is trivial and G is not.

Example 3.01. As a further illustration we list all the subgroups of the symmetric group S_3 on $\{1, 2, 3\}$. There are, in addition to S_3 and 1:

$$\{1,\ (123),\ (132)\},$$
$$\{1,\ (12)\},$$
$$\{1,\ (13)\},$$
$$\{1,\ (23)\}.$$

These are, it will be found, the only complexes closed under multiplication; and it may be verified that they are in fact subgroups.

The positive rationals with multiplication are *not* regarded as a subgroup of the additive rationals, since they do not form a group under addition of rationals.

Since we shall find important complexes which themselves are elements of groups, some consideration must be given to complex multiplication and its properties.

DEFINITION. The *product* CD of the complexes C, D (in that order) in a given group is $\{cd : c \in C,\ d \in D\}$.

Example 3.02. As an illustration we calculate $C_1 C_2$ and $C_1 C_3$ where C_i are the following complexes in S_3:

$$C_1 = \{(12),\ (13)\}, \quad C_2 = \{(12),\ (23)\}, \quad C_3 = \{(13),\ (23)\}.$$

We find that

$$\begin{aligned} C_1 C_2 &= \{(12)(12),\ (12)(23),\ (13)(12),\ (13)(23)\} \\ &= \{1,\ (123),\ (132)\}, \\ C_1 C_3 &= \{(12)(13),\ (12)(23),\ (13)(13),\ (13)(23)\} \\ &= \{1,\ (123),\ (132)\}. \end{aligned}$$

The meaning of equations like $CD = DC$ should be carefully noted; it is *not* meant that each element of C commutes with each element of D. The meaning of $CD \subseteq DC$ is that for each c in C and each d in D there is an element c' in C and an element d' in D for which $cd = d'c'$. It will be found that $C_1 C_2 = C_2 C_1$

in our example above, although (13) in C_1 does not commute with (23) in C_2.

The fact that a given complex C of a group G is closed *under the multiplication in the group* G is neatly expressed by $C^2 \subseteq C$. Note that if C contains 1 then both $D \subseteq CD$ and $D \subseteq DC$ hold for all complexes D; so if C is multiplicatively closed and contains 1 then $C^2 = C$. In particular $S^2 = S$ whenever S is a subgroup of G.

THEOREM 3.03. *Multiplication of complexes in a given group is associative.*

Proof. Take complexes C_1, C_2, C_3 in a given group. We are required to prove that

$$C_1(C_2 C_3) = (C_1 C_2)C_3.$$

Take $x \in C_1(C_2 C_3)$, then $x = c_1(c_2 c_3)$ where $c_i \in C_i$ for $i = 1, 2, 3$. By the associative law in the group,

$$c_1(c_2 c_3) = (c_1 c_2)c_3.$$

Hence $x \in (C_1 C_2)C_3$, that is,

$$C_1(C_2 C_3) \subseteq (C_1 C_2)C_3.$$

But since a very similar proof shows that

$$C_1(C_2 C_3) \supseteq (C_1 C_2)C_3,$$

we have $C_1(C_2 C_3) = (C_1 C_2)C_3$, as required. □

It follows that a multiplicatively closed set of complexes in a given group forms a semigroup. Such a set of complexes will not usually form a group. The fact that the remaining group axioms cannot all hold in general is shown by Example 3.02, in which the complexes C_1, C_2, C_3 of S_3 are such that $C_1 C_2 = C_1 C_3$ and $C_2 \neq C_3$.

It is now possible to use expressions like $C_1 C_2 C_3$ without ambiguity.

There are special sets of complexes in the group G that do form groups under complex multiplication. One is the set $\{G\}$ consisting of the one complex G. Another is the set of one-element complexes of G. Any complex C in a group G with the property that $C^2 = C$ gives a one-element group $\{C\}$. A less trivial example may be constructed in S_3. Let

$$C = \{1, (123), (132)\},$$
$$D = \{(12), (13), (23)\}.$$

A little calculation gives

$$C^2 = D^2 = C, \quad CD = DC = D.$$

It follows that $\{C, D\}$ is a group under complex multiplication, the identity element being C.

We mention a theorem that relates the operation of complex multiplication with the set operations of taking unions and intersections.

THEOREM 3.04. *If C_1, C_2, C_3 are complexes of a given group then*

(i) $C_1(C_2 \cup C_3) = C_1 C_2 \cup C_1 C_3$;

(ii) $(C_2 \cup C_3)C_1 = C_2 C_1 \cup C_3 C_1$;

(iii) $C_1(C_2 \cap C_3) \subseteq C_1 C_2 \cap C_1 C_3$;

(iv) $(C_2 \cap C_3)C_1 \subseteq C_2 C_1 \cap C_3 C_1$.

Proof. In view of the small use we intend to make of this theorem we offer only a specimen by way of proof. Take (iii). Let $x \in C_1(C_2 \cap C_3)$, so that $x = yz$ where $y \in C_1$, $z \in C_2$, $z \in C_3$. It follows that $yz \in C_1 C_2$ and $yz \in C_1 C_3$, and so $x = yz \in C_1 C_2 \cap C_1 C_3$; that is $C_1(C_2 \cap C_3) \subseteq C_1 C_2 \cap C_1 C_3$, as required. □

Our next theme, a criterion for a complex to be a subgroup, needs some preparation.

THEOREM 3.05. *If S is a subgroup of the group G, then the identity element of G is the same as that of S, and the inverse of each element of S is the same in G as in S.*

Proof. In any group the identity element is the only solution to the equation $x^2 = x$. Since the multiplications in S and G are the same and since the identity element of S is an element of G for which $x^2 = x$, the identity elements of S and G coincide. For each x in S the inverse x^{-1} in S is clearly the inverse in G, therefore; so inverses in S and G coincide. □

If S is a subgroup of G then we could of course construct a homomorphism (namely the inclusion mapping) from S, considered as a group in its own right, into G. In that case Theorems 2.13 and 3.05 become very similar.

THEOREM 3.06. *A non-empty complex C of a group G is a subgroup of G if and only if xy^{-1} lies in C for each x, y in C.*

Proof. It is clear that if C is a subgroup containing x and y then y^{-1} and so xy^{-1} lie in C. Note that any subgroup contains an identity element (in view of the definition of a group) and is therefore non-empty.

Suppose conversely that C is a non-empty complex containing xy^{-1} whenever it contains x and y. In particular, for each x in C, the element $xx^{-1} = 1$ lies in C. Taking $x = 1$, we next have $1y^{-1} = y^{-1}$ lying in C. And then, for each x and y in C, we have $x(y^{-1})^{-1} = xy$ also a member of C. Therefore C is closed under the multiplication of G, which we are, of course, using as the multiplication in C also.

The associative law for multiplication of a, b, c in C holds because a, b, c are elements of G, in which the associative law certainly holds. We have shown that if C is non-empty then C contains 1, which also acts as an identity element in C. Finally x^{-1}, the inverse of x in G, lies in C whenever x does, and serves as a required inverse in C. Therefore C is a subgroup of G. □

Note that if we define C^{-1} to be $\{x : x^{-1} \in C\}$ then the theorem may be stated succinctly in this way: $C \leqslant G$ if and only if $C \neq \emptyset$, $C \subseteq G$, and $CC^{-1} \subseteq C$.

This theorem makes it easy to verify the following statement: if g is any element of a group G then the complex

$$\{g^n : n = 0, \pm 1, ...\}$$

is a subgroup of G. We thus have many subgroups, one determined by each element g of G, though these need not all be distinct. They are abelian, and we shall have more to say about their nature later on.

Two lines of thought lead to our next batch of results about relations between complexes and subgroups. In the first place one may ask whether the usual set operations can be performed on subgroups with pleasing results; in other words, is the intersection or the union of a set of subgroups again a subgroup? The union of subgroups certainly need not be a subgroup—for example, $\{1, (12)\} \cup \{1, (13)\}$ is not a subgroup of the symmetric group on $\{1, 2, 3\}$. This focuses attention on another problem—what can be said about subgroups containing a given complex? When this is answered satisfactorily we can consider subgroups containing the union of given subgroups, which is itself a complex, and so hope to find a group-theoretic analogue of set-theoretic union.

THEOREM 3.07. *Let $\{S_\lambda : \lambda \in \Lambda\}$ be a set of subgroups of the group G, where Λ is some index set. Then $\bigcap_{\lambda \in \Lambda} S_\lambda$ is a subgroup of G.*

Proof. Let S denote $\bigcap_{\lambda \in \Lambda} S_\lambda$. Note that S is non-empty since $1 \in S_\lambda$ for each λ and so $1 \in S$. Take arbitrary elements x, y in S. Since x, y belong to each subgroup S_λ and so y^{-1} belongs to each S_λ, we find that xy^{-1} belongs to each S_λ. Hence xy^{-1} is an element of S. By Theorem 3.06 it follows that S is a subgroup of G. □

Example 3.08. The additive group Z provides an illustration of this. It may be verified that the complexes $\{k2^i : k = 0, \pm 1, \ldots\}$ form subgroups, for $i = 1, 2, \ldots$. Their intersection is $\{0\}$, again a subgroup.

If S_1, S_2 are subgroups of a group G then it rarely happens that $S_1 \cup S_2$ is a subgroup of G. In fact this happens if and only if one of S_1, S_2 is contained in the other. We prove briefly part of this statement. Suppose that G has an element a which lies in S_1 but not in S_2, and take any element x in S_2. If ax lies in S_2 then a lies in S_2, a contradiction; therefore ax lies in S_1, which implies that x lies in S_1. We have now shown that S_2 is a subset of S_1, and the required result follows.

It is also exceptional to have the complex $S_1 S_2$ a subgroup when S_1, S_2 are subgroups of G. Examples to substantiate this assertion may again be found in the symmetric group on $\{1, 2, 3\}$.

However, there is at least one subgroup of G that contains a given complex C, for G itself is such a subgroup. Further, the intersection of all subgroups of G containing C is another such subgroup, for this intersection clearly contains C and by Theorem 3.07 is itself a subgroup. Clearly the intersection is the smallest subgroup containing C, in the sense that it is contained in any subgroup that contains C. This leads to a definition and a theorem.

DEFINITION. The subgroup *generated* by a non-empty complex C in the group G is the intersection of all subgroups of G containing C. It will be denoted by $\mathrm{gp}\{C\}$. When C is finite,

$$C = \{c_1, \ldots, c_n\},$$

we shall write $\mathrm{gp}\{C\}$ as $\mathrm{gp}\{c_1, \ldots, c_n\}$, and we shall say that this subgroup is *finitely generated*. There are finitely generated groups which are not finite; the infinite group Z is generated by the integer 1, for instance.

Example 3.09. Consider the group Z_p^∞ (Example 1.09) and its subgroup generated by C where

$$C = \{c_1, \ldots, c_n\},$$

$$c_m = \cos\frac{2\pi}{p^m} + i\sin\frac{2\pi}{p^m} \quad (1 \leqslant m \leqslant n),$$

for some fixed positive integer n. De Moivre's theorem gives $c_m^p = c_{m-1}$ for each $m > 1$; thus $c_{m-1} \in \mathrm{gp}\{c_m\}$, and by induction $c_m \in \mathrm{gp}\{c_n\}$ for each m. It follows that $C \subseteq \mathrm{gp}\{c_n\}$. Then $\mathrm{gp}\{C\}$ is contained in $\mathrm{gp}\{c_n\}$, by definition of the former. It should now be clear that $\mathrm{gp}\{C\} = \mathrm{gp}\{c_n\}$.

The above definition does not provide a very satisfactory *practical* method of determining the subgroup generated by a complex. If we are asked to find the subgroup of S_3 generated by $\{(12), (13)\}$, the natural approach is to add to this complex elements found by multiplication and inversion until a subgroup is formed. Thus

$$(12)(12) = 1,$$
$$(12)(13) = (123),$$
$$(13)(12) = (132),$$
$$(12)(13)(12) = (23);$$

so we have proved that $S_3 = \text{gp}\{(12), (13)\}$. We next make this idea precise.

THEOREM 3.10. *The subgroup generated by the non-empty complex C of the group G is the complex containing 1 and all elements of the form $c_1 c_2 \ldots c_n$ for each $n \geqslant 1$, where each c_i lies in $C \cup C^{-1}$.*

Proof. By C^{-1} we mean $\{x : x^{-1} \in C\}$.

Let $C^* = \{1\} \cup \{c_1 c_2 \ldots c_n : n \geqslant 1, \ \ c_i \in C \cup C^{-1}\}$. Clearly $C \subseteq C^*$, for each element of C has the form required of elements in C^* with $n = 1$. Further C^* is a subgroup, for it is evidently a non-empty complex with the property that if x, y lie in C^* then so do y^{-1} and xy^{-1}. Therefore C^* is a subgroup containing C, and by the definition of $\text{gp}\{C\}$ we have

$$\text{gp}\{C\} \subseteq C^*.$$

Now take an arbitrary element x of C^*. If $x = 1$ then certainly x lies in $\text{gp}\{C\}$. If $x = c_1 \ldots c_n$ with each c_i in $C \cup C^{-1}$ then the fact that $\text{gp}\{C\}$ is closed under the operations of inversion and multiplication shows that $x \in \text{gp}\{C\}$ again. (To be strict, induction on n is required to show that c_1, $c_1 c_2, \ldots, c_1 c_2 \ldots c_n$ all lie in $\text{gp}\{C\}$.) Hence

$$C^* \subseteq \text{gp}\{C\}.$$

It follows that $\text{gp}\{C\} = C^*$, as required. □

We ought formally to note that our labours have shown this: the group-theoretic analogue of $S_1 \cup S_2$, where S_1 and S_2 are subgroups of a given group, is $\text{gp}\{S_1, S_2\}$.

If we look at the idea of generators for a group from another angle, a further problem arises. Given a group G, can we find a complex C for which $\mathrm{gp}\{C\} = G$? We certainly can, merely by taking $C = G$; that is, by saying that G is generated by the set of all its elements. But in practice we usually want generating sets with special properties, such as being finite or not containing a proper subset that generates G (though neither of these properties can generally be arranged for). We shall do no more than note that such problems are difficult or impossible in general, and indicate (without proof) two of the several possible generating sets for S_3:

$$\{(12),\ (13)\},$$

$$\{(23),\ (123)\}.$$

The central part in our study of complexes has been played by subgroups, and it is now time to begin using an additional tool to investigate group structure.

THEOREM 3.11. *If the relation $x \sim y$ is defined between the pair of elements x, y of the group G to mean that xy^{-1} lies in the fixed subgroup S, then $x \sim y$ is an equivalence relation on G.*

Proof. (i) Since $xx^{-1} = 1$ lies in the subgroup S, we have $x \sim x$.

(ii) Suppose $x \sim y$; that is, xy^{-1} lies in S. Then $(xy^{-1})^{-1} = yx^{-1}$ lies in S, and so $y \sim x$.

(iii) Suppose $x \sim y$ and $y \sim z$, and so xy^{-1} and yz^{-1} lie in S. Then $xz^{-1} = (xy^{-1})(yz^{-1})$ lies in S, and we have $x \sim z$.

Since the relation is therefore reflexive, symmetric, and transitive, it is an equivalence relation. □

DEFINITION. A *right coset* of the subgroup S in the group G is an equivalence class resulting from the equivalence relation defined above.

Since this set of right cosets provides a partition of G, it is natural to seek an explicit description of right cosets. The equivalence class containing a is $\{x : xa^{-1} \in S\}$, which may be written as $\{x : x \in Sa\}$ or simply as Sa, if the latter is taken to mean $S\{a\}$ or $\{x : x = sa$ for some s in $S\}$.

Left cosets may be defined in a similar but obvious way. One starts with the relation $x \sim y$ defined as $x^{-1}y \in S$, proves this to be an equivalence relation, defines a left coset as an equivalence class, and can show that the coset containing a is aS with suitable interpretation.

We shall see in Chapter 4 that certain sets of cosets in any group themselves form a group under complex multiplication.

Let us calculate the right and left cosets of the subgroup S generated by (12) in the symmetric group on $\{1, 2, 3\}$. The right cosets include the subgroup itself, $\{1, (12)\}$; next, $S(13)$ is $\{(13), (123)\}$; and $S(23)$ is $\{(23), (132)\}$. We have arrived at a partition of the group, and so have a complete set of right cosets. It will similarly be found that the left cosets are

$$S = \{1, (12)\},$$
$$(13)S = \{(13), (132)\},$$
$$(23)S = \{(23), (123)\}.$$

Note that $S(13) = S(123)$; any element chosen from a coset determines that coset.

THEOREM 3.12. *Let S be a subgroup, and a and b be elements, of the group G. Then $Sa = Sb$ if and only if a and b lie in the same right coset of S in G.*

Proof. Suppose, in the first place, that $Sa = Sb$. This implies that $a = xb$ for a suitable element x in S; and so $ab^{-1} \in S$. By definition of a right coset, a and b lie in the same right coset of S in G.

To prove the converse implication, suppose that a and b lie in the same right coset. We find that $ab^{-1} \in S$, and $a = xb$ for some x in S. It follows that $Sa = Sxb$. Therefore we have to prove no more than $Sx = S$. Because S is a subgroup we have $Sx \subseteq S$; and if $y \in S$ then $yx^{-1} \in S$, as in Theorem 3.06, and so $y \in Sx$, proving that $S \subseteq Sx$. Therefore $Sx = S$. □

COROLLARY 3.13. *Let S be a subgroup, and a an element, of the group G. Then $Sa = S$ if and only if a lies in S.*

Proof. Take $b = 1$ in the theorem above. □

Another aspect of Theorem 3.12 should be carefully noted. Suppose Sa is a right coset of S in G, and suppose that an *arbitrary* element z is chosen in Sa; then the cosets Sa and Sz coincide. It is sometimes useful to be able to rename cosets in this way.

We remark that everything above, after the statement of Theorem 3.12, can be rewritten for *left* cosets.

DEFINITION. A *right transversal* of the group G with respect to the subgroup S is any complex of G containing one and only one element from each right coset of S in G. Left transversals are defined similarly. Sometimes a right (or left) transversal is called a *complete set of right (left) coset representatives.*

In our example above, one right transversal is $\{1, (13), (23)\}$, which is also a left transversal. Another is $\{1, (13), (132)\}$, but this is not a left transversal.

The final aim of this chapter is to study some important arithmetical conditions satisfied by finite groups. We need some definitions; these will also be very useful in the study of infinite groups.

DEFINITION. The *order of a group* is the number of elements it contains. The order of G will be denoted by $|G|$.

The order may be a positive integer or may be infinite. For example, the order of S_n is $n!$, while the orders of Z, Q, R are all infinite.

DEFINITION. The *order of an element* g in the group G is the least positive integer n such that $g^n = 1$, if n exists; g has infinite order if such an integer n does not exist. The order of g will be denoted by $|g|$.

For instance, the order of the element (123) in S_3 is 3; the order of the number 2 in the multiplicative group of positive rationals is infinite. An element of a group G has order 1 if and only if it is the identity element, clearly.

DEFINITION. A *periodic* group is a group in which every element has finite order.

Every finite group is clearly periodic, but there are also infinite examples. The reader will find it easy to show that the permutation group (on the positive integers) generated by $\{(12), (34), (56),...\}$ is infinite and periodic.

DEFINITION. A *torsionfree* group is a group in which every element, except the identity, has infinite order.

It is easy to verify that Z, Q, R are all torsionfree.

The next result shows that the two meanings given to 'order' by our definitions are sensibly related.

THEOREM 3.14. *Let g be an element of a fixed group. Then*

(i) $g^{m_1} = g^{m_2}$ *if and only if* $m_1 \equiv m_2$ *modulo* $|g|$;

(ii) *the order of g equals the order of the subgroup generated by g.*

Proof. (i) If g has infinite order then we interpret the congruence to mean that $m_1 = m_2$. Since $g^{m_1} = g^{m_2}$ is equivalent to $g^{m_1-m_2} = 1$, and $g^{m_1-m_2} = 1$ is equivalent to $m_1 - m_2 = 0$ in this case, the required statement is proved.

Next suppose that g has finite order n. Let $g^{m_1} = g^{m_2}$, so that $g^{m_1-m_2} = 1$, and put $m = m_1 - m_2$. By the division algorithm† there are integers u, v for which $m = un+v$ and $0 \leqslant v < n$. Therefore, by Theorems 2.08 and 2.09,

$$1 = g^m = g^{un+v} = (g^n)^u g^v.$$

Since g has order n we conclude that $g^v = 1$. But since n is the *minimal* positive integer for which $g^n = 1$, and $0 \leqslant v < n$, we have $v = 0$; that is, n divides m, and so $m_1 \equiv m_2$ modulo n.

Now let $m_1 \equiv m_2$ modulo n, with n as above the order of g. Then $m_1 = m_2+un$ for some integer u, and

$$g^{m_1} = g^{m_2+un} = g^{m_2}(g^n)^u = g^{m_2};$$

this gives $g^{m_1} = g^{m_2}$ as required.

† See the Appendix.

(ii) It is now clear that the number of distinct powers of g equals the order of the element g. That is to say, the order of $\text{gp}\{g\}$ equals the order of g. □

A special case of the theorem is worth mentioning; this is the fact that if g has finite order n then $g^m = 1$ if and only if m is a multiple of n.

Theorem 3.14 is often used to prove that two elements have the same order. Take inverse elements a and a^{-1} in any group, for instance. If a has order n then $a^n = 1$, and so $(a^n)^{-1} = 1$, and so $(a^{-1})^n = 1$; all we can conclude at this stage is that a^{-1} has finite order *dividing* n. But another application of the same argument shows that the order of a divides the order of a^{-1}, and it follows that they are equal. It should be clear too that a has infinite order if and only if a^{-1} has infinite order.

The theorem that follows is fundamental in the theory of finite groups. It is due to Lagrange.

THEOREM 3.15. *The order of a subgroup of a finite group divides the order of the group.*

Proof. Let S be a subgroup of the finite group G. We have seen that the right cosets of S in G give a partition of G. But each right coset contains n elements, if n is the order of S; for there is a one–one correspondence between S and Sa. (This may be set up by making xa in Sa correspond to x in S; if $xa = ya$ then clearly $x = y$.) Thus if there are k right cosets then the order of G is kn. □

COROLLARY 3.16. *The order of an element of a finite group divides the order of the group.*

Proof. We need remark only that the order of an element equals the order of the subgroup that it generates, by Theorem 3.14. □

The following converse of Lagrange's theorem is false: a group of finite order n contains a subgroup of order m for each factor m of n. It will be found, for instance, that the alternating group A_4 on four symbols has no subgroup of order 6.

Another consequence of Lagrange's theorem is that the number of left cosets of S in G is the same as the number of right cosets when G is finite. A little preparation is needed to generalize this satisfactorily to infinite groups.

THEOREM 3.17. *If T is a right transversal for the group G with respect to the subgroup S then T^{-1} is a left transversal in the same situation.*

Proof. By T^{-1} we mean, as usual, $\{x : x^{-1} \in T\}$.

Let x be an arbitrary element of G. Then x^{-1} may be expressed as st, with $s \in S$, $t = T$. Therefore $x = t^{-1}s^{-1} \in t^{-1}S$, since S is a subgroup. Each element of G then belongs to some left coset $t^{-1}S$ where $t \in T$. Now suppose that $t_1^{-1}S = t_2^{-1}S$ where t_1, t_2 lie in T. Then $t_1^{-1} = t_2^{-1}s$ for some $s \in S$, and $t_1 t_2^{-1} \in S$. This implies that $t_1 = t_2$ because T was a right transversal. So we have a partition of G into left cosets, and T^{-1} is a left transversal. □

COROLLARY 3.18. *The right cosets of S in G are in one–one correspondence with the left cosets of S in G.* □

DEFINITION. The *index* of the subgroup S in G is the number of (right or left) cosets of S in G. This index will be denoted by $|G : S|$.

Note that the last result shows that in calculating an index it makes no difference whether we work with right cosets or left cosets. If G has finite order then its order is the product of the order of S and the index of S in G. Indeed the reader should have little difficulty in proving a more general fact: if G has a subgroup H which in turn has a subgroup K, then

$$|G : K| = |G : H||H : K|.$$

Some facts about subgroups with finite index will now concern us. These facts may be of no great significance in themselves, but they are interesting exercises on the previous theorems and they will be very helpful in the later chapters.

LEMMA 3.19. *Let A and B be subgroups of the group G. If x and y are elements of G for which $Ax \cap By$ is non-empty, then $Ax \cap By$ is a right coset of $A \cap B$ in G.*

Proof. Let $z \in Ax \cap By$. Since $z \in Ax$ we have $Ax = Az$, by Theorem 3.12 and a remark following; similarly $By = Bz$. Therefore $Ax \cap By = Az \cap Bz$. It is easily verified that $Az \cap Bz \supseteq (A \cap B)z$, and we prove the reverse inclusion as follows. If $z_0 \in Az \cap Bz$ then $z_0 \in Az$ and $z_0 \in Bz$. Therefore

$$z_0 = az, \quad z_0 = bz$$

for suitable a in A and b in B. Clearly $a = b$, and clearly this element lies in $A \cap B$. Therefore $z_0 \in (A \cap B)z$, as required. It follows that

$$Ax \cap By = Az \cap Bz = (A \cap B)z. \square$$

THEOREM 3.20. *Let A and B be subgroups of the group G with finite indices m, n respectively. Then*

(i) *$A \cap B$ has finite index in G, bounded by mn;*
(ii) *if m and n are coprime then $A \cap B$ has index mn in G.*

Proof. (i) There are m right cosets of the form Ax and n of the form By. It follows that there are at most mn non-empty intersections $Ax \cap By$. The lemma above shows that the set of such non-empty intersections is just the set of right cosets of $A \cap B$ in G. Therefore $A \cap B$ has finite index in G, and mn is an upper bound for this index.

(ii) When m and n are coprime we denote the index of $A \cap B$ in G by k. We know that $k \leqslant mn$. Since

$$A \cap B \leqslant A \leqslant G, \quad A \cap B \leqslant B \leqslant G,$$

the following equalities are obvious:

$$|G:A \cap B| = |G:A||A:A \cap B|,$$
$$|G:A \cap B| = |G:B||B:A \cap B|.$$

The former implies that m divides k, and the latter that n divides k. We conclude that mn divides k because m and n are coprime. This fact together with $k \leqslant mn$ shows that $k = mn$. □

Note that in case (ii) of the above theorem $Ax \cap By$ is never empty. The reader might like to prove that in general $Ax \cap By$ may well be empty, implying that the coprime condition in case (ii) cannot be abandoned.

The determination of all subgroups of a given group G, or even some description of their properties, is in general an unreasonably difficult problem. This is so even when G is generated by two elements. But the case when G is generated by one element is surprisingly easy to discuss.

DEFINITION. A *cyclic* group is a group generated by one of its elements.

Thus if g generates G then $G = \{g^n : n = 0, \pm 1, \ldots\}$. Clearly all cyclic groups are abelian. There are cyclic groups of every finite order and of infinite order, for Z_n and Z are cyclic. Each element of an arbitrary group G clearly lies in, indeed generates, a cyclic subgroup of G.

THEOREM 3.21. *Every subgroup of a cyclic group is cyclic.*

Proof. Let S be a subgroup of the cyclic group G with generator g. The elements of S are certain powers of g. If S is not the subgroup containing only one element then $g^\lambda \in S$ where $\lambda \neq 0$; since S is a subgroup we may take $\lambda > 0$. Let λ be the *minimal* positive integer such that $g^\lambda \in S$, and let g^μ be an arbitrary element of S. By the division algorithm $\mu = u\lambda + v$ where u, v are integers and $0 \leqslant v < \lambda$. Therefore

$$g^v = g^\mu (g^\lambda)^{-u},$$

and it follows that $g^v \in S$. The minimal property of λ now shows that $v = 0$. Therefore λ divides μ, and g^μ is a power of g^λ; this proves that S is cyclic with generator g^λ. □

There is usually more than one generator of a cyclic group, and we offer some examples of this situation in place of an exhaustive discussion. The additive group of integers is generated by 1 or by -1, but by no other integer. The group Z_n is generated by [1]. The group Z_{12} is generated by [1] or [5] or [7] or [11]—it is clearly generated by [1] and hence by the others as

$$5[5] = [1], \quad 7[7] = [1], \quad 11[11] = [1].$$

However, [3] is *not* a generator, for there is no integer k for which $k[3] = [1]$; if there were, then $3k-1$ would be divisible by 12 and so by 3, a contradiction.

Further properties of cyclic groups are presented among the problems which follow.

Problems

1. Show that the residue classes $[2n+1]$ modulo 16, where $0 \leqslant n \leqslant 7$, form a multiplicative group, and find all its subgroups.

2. Find the order of the subgroup generated by the following complex of the symmetric group S_6:

$$\{(1234),\ (14)(32),\ (56)\}.$$

3. Prove that a group is finite if and only if it contains only finitely many subgroups.

4. Show that the alternating group on five symbols has no subgroup of index 2, 3, or 4, and find a subgroup of index 5.

5. Let $\mathscr{D}$ be the multiplicative group generated by the matrices A, B, where

$$A = \begin{pmatrix} 1 & 0 \\ 0 & -1 \end{pmatrix}, \qquad B = \begin{pmatrix} 1 & 1 \\ 0 & -1 \end{pmatrix}.$$

(i) Show that A and B have order 2.

(ii) Prove that every element of $\mathscr{D}$ except I can be written in the form $C_1 \dots C_n$ for some $n \geqslant 1$, with each $C_i = A$ or B, and $C_i \neq C_{i+1}$ for $i = 1, 2, \dots, n-1$. Prove that no such element is the identity, and deduce that AB has infinite order.

6. Find examples to prove that each of the following statements about the group G is false.

(i) If every element of G has order dividing n then G has order dividing n.

(ii) If G is finite and every proper subgroup is cyclic then G is cyclic.

(iii) If G has finite order divisible by n then G contains an element of order n.

(iv) If G contains a complex C satisfying the conditions $C^2 = C$ and $1 \in C$, then C is a subgroup of G.
(v) If S_1, S_2, S_3 are subgroups of G then $S_1(S_2 \cap S_3) = S_1 S_2 \cap S_1 S_3$.

7. Let C be a non-empty complex of the periodic group G. Prove that the following conditions on C are equivalent:

(i) $CC^{-1} \subseteq C$;
(ii) $C^2 = C$;
(iii) $C^2 \subseteq C$.

8. Which of the following groups are cyclic? Which are finitely generated?

(i) The multiplicative group $\{[n]:n \text{ is not divisible by } 5\}$ of residue classes modulo 25.
(ii) The additive group of rationals.
(iii) The additive group of reals.
(iv) The multiplicative group of positive rationals.
(v) The multiplicative group Z_p^∞ of complex numbers (Example 1.09)

9. Let S be a non-empty subset of the group G, and define the relation $x \sim y$ between elements x, y to mean that $xy^{-1} \in S$. Prove that if this is an equivalence relation then S is a subgroup of G.

Another relation in G is obtained by defining $x \sim y$ to mean that $Sx = Sy$. Show that although this is an equivalence relation there are cases in which S is not a subgroup of G.

10. Let a, b denote the permutations (12), $(12 \ldots n)$ respectively in the symmetric group S_n on n symbols; let $n \geqslant 2$. Prove that

$$b^{-i}ab^i = (i+1, i+2) \quad \text{for } i = 0, 1, \ldots, n-2.$$

By showing that *every* transposition can be expressed as a suitable product of such transpositions $(i+1, i+2)$, prove that $S_n = \text{gp}\{a, b\}$.

*11. Let $\mathscr{G}$ be the multiplicative group generated by the matrices

$$A = \begin{pmatrix} 1 & 1 \\ 0 & 1 \end{pmatrix}, \qquad B = \begin{pmatrix} 2 & 0 \\ 0 & 1 \end{pmatrix}.$$

Prove that $BAB^{-1} = A^2$, and deduce that every element of $\mathscr{G}$ may be written in the form $B^\alpha A^\beta B^\gamma$ for suitable integers α, β, γ.

*12. Let G be a finite cyclic group of order n. Prove that if m is any factor of n then G contains one and only one subgroup of order m.

Deduce that a group has no proper subgroup if and only if its order is prime or 1.

*13. The complex C_r, where r is *rational*, in the (additive) group R of reals, is defined by

$$C_r = \{x : x > r\}.$$

Prove that $C_r C_s = C_{r+s}$, where r, s are rational, and complex multiplication is used. Deduce that the set $\{C_r : r \in R\}$ forms a group under complex multiplication.

**14. The finite group G contains a set of complexes that forms a group under complex multiplication. Prove that each complex contains the same number of elements of G. Prove further that, if the complex C_1 acts as the unit element of the complex group and x lies in the complex C, then $C = xC_1 = C_1 x$.

Deduce that if G is cyclic then the complex group is cyclic.

4

Normal subgroups and factor groups

In this chapter we continue to assemble the stock of elementary concepts and tools necessary for deep penetration into the theory of groups. The key idea is that of a normal subgroup, whose immediate justification is that it enables us to solve the following obviously important problem: given a group G, what groups can occur as homomorphic images of G? That is to say: what description can be found for each group H for which there exists a homomorphism from G onto H?

Definition. The subgroup N of the group G is *normal* if $xN = Nx$ for each element x of G. The fact that N is a normal subgroup of G is expressed symbolically thus: $N \trianglelefteq G$. If $N \trianglelefteq G$ and $N \neq G$, write $N \lhd G$.

It should be noted with considerable care that the equation $xN = Nx$ does *not* assert that every element of N commutes with x. What is intended is that $\{x\}$ and N shall commute as complexes. In other words, the left coset xN is identical with the right coset Nx. This condition is equivalent to $xN \subseteq Nx$ and $xN \supseteq Nx$ simultaneously; the former of this pair means that, given y in N, an element z (depending on y) exists in N such that $xy = zx$, and the latter is to be similarly interpreted.

An immediate consequence of the definition is the fact that $SN = NS$ for any subgroup S of the group G. (Indeed this equation holds if S denotes any complex of G.)

It is often helpful to use the fact that the subgroup N is normal in G if $x^{-1}Nx = N$ for all x in G; this is another one-step corollary of the above definition.

The normal subgroups 1 and G are present in every group G, for which reason they are often called the trivial normal subgroups. It should be clear that *every* subgroup of any abelian group is normal. Some less trivial examples follow.

Example 4.01. Consider the symmetric group S_3 on $\{1, 2, 3\}$ once again. We assert that $\{1, (123), (132)\}$ is a normal subgroup. The non-trivial part of the proof of this is to verify the following statements:

$$(12)\{1, (123), (132)\} = \{(12), (13), (23)\} = \{1, (132), (123)\}(12),$$

$$(13)\{1, (123), (132)\} = \{(13), (23), (12)\} = \{1, (132), (123)\}(13),$$

$$(23)\{1, (123), (132)\} = \{(23), (12), (13)\} = \{1, (132), (123)\}(23).$$

There is a more conceptual proof that the subgroup in question is normal, and it can be put in a quite general setting. Suppose that G is any group with a subgroup S of index 2. Take any element x in G but not in S. Then $S \cap Sx = \emptyset$ while $G = \{S, Sx\}$ by properties of cosets, and it follows that the elements of Sx are precisely the elements of G which are *not* in S. But a quite similar argument shows that these elements are just the set xS. Therefore $Sx = xS$. That is to say, any subgroup of index 2 is normal in G. In particular, the alternating group A_n is normal in the symmetric group S_n because it has the required index. Our example above was the case $n = 3$.

It may be easily shown that S_3 has no normal subgroup of index 3.

Example 4.02. Consider $\mathscr{M}_2$, the multiplicative group of all non-singular 2×2 matrices with real entries. Let $\mathscr{N}$ be the subset of all matrices of unit determinant. It should be clear that $\mathscr{N}$ is a subgroup of $\mathscr{M}_2$, and we give a few words of proof to indicate that it is normal. Let A be a matrix in $\mathscr{N}$, and let X be any matrix in $\mathscr{M}_2$; since

$$|X^{-1}AX| = |X|^{-1}|A||X| = |A| = 1$$

it follows that $X^{-1}AX$ is in $\mathscr{N}$. Hence $X^{-1}\mathscr{N}X \subseteq \mathscr{N}$. This argument, with X replaced by X^{-1}, gives $X\mathscr{N}X^{-1} \subseteq \mathscr{N}$. We now have the relations

$$X^{-1}\mathscr{N}X \subseteq \mathscr{N}, \qquad \mathscr{N} \subseteq X^{-1}\mathscr{N}X;$$

it follows that $X^{-1}\mathscr{N}X = \mathscr{N}$ for all X in $\mathscr{M}_2$. Hence $\mathscr{N}$ is a normal subgroup of $\mathscr{M}_2$.

It is perhaps worth stating and noting carefully that, if the subgroup S of the group G satisfies the condition $x^{-1}Sx \subseteq S$ for *all* x in G, then S is normal. The useful trick (exhibited in Example 4.02) of replacing x by x^{-1} will establish this assertion.

DEFINITION. The *kernel* K of the homomorphism ϕ from the group G into the group H is the set $\{x : x \in G, x\phi = 1\}$.

As an illustration we indicate the kernel of the homomorphism ϕ from $\mathscr{M}_2$ to R^* described in Example 2.11. If $A \in \mathscr{M}_2$ then $A\phi = |A|$; therefore the kernel consists of those matrices A which have determinant 1. As shown above in Example 4.02, such matrices form a normal subgroup of $\mathscr{M}_2$.

THEOREM 4.03. *The kernel of a homomorphism from a group G is a normal subgroup of G.*

Proof. Let ϕ be a homomorphism from the group G into the group H. We show first that the kernel K is a subgroup. As $1 \in K$, clearly, K is non-empty. In order to apply the usual subgroup criterion we take arbitrary elements x, y in K. Thus $x\phi = 1$, $y\phi = 1$. Then

$$(xy^{-1})\phi = (x\phi)(y\phi)^{-1} = 1,$$

since ϕ is a homomorphism. The consequent fact that $xy^{-1} \in K$ implies that K is a subgroup of G, by Theorem 3.06.

Next we have to show that K is a normal subgroup. Let k, g be arbitrary elements of K, G respectively. Then, since ϕ is a homomorphism,

$$(g^{-1}kg)\phi = (g\phi)^{-1}(k\phi)(g\phi) = (g\phi)^{-1}1(g\phi) = 1.$$

Thus $g^{-1}kg \in K$. We have proved that $g^{-1}Kg \subseteq K$ for all g in G, and this suffices to establish that K is a normal subgroup of G. □

We now investigate cosets of a normal subgroup.

THEOREM 4.04. *The right cosets of the subgroup S in the group G form a group under complex multiplication if and only if S is normal in G.*

Proof. Suppose S is normal in G. What is the product of the cosets Sx and Sy? It is the complex $SxSy$. But note that the relation $xS = Sx$ allows us to write this as S^2xy, or as Sxy (by the subgroup property of S). Therefore if S is normal then we have

$$SxSy = Sxy$$

for all x and y in G.

Verification of the group axioms for the set of cosets is now easy. Proof of associativity is standard. The rest follows from the above equation—closure is obvious, the identity element is the coset S, and the inverse of the coset Sx is Sx^{-1}. Thus the cosets of S in G form a complex group.

To prove the theorem fully we have to deduce normality of S in G from the fact that the cosets of S form a complex group. This fact, though not essential to the development of the subject, is none the less of some interest.

We are supposing that corresponding to each pair x, y of elements in G there is an element z for which

$$SxSy = Sz.$$

Since 1 lies in S there is an element s in S for which

$$xy = 1x1y = sz.$$

Therefore

$$SxSy = Sxy,$$

since $Ss = S$, and so $Ssz = Sz$. Take $y = 1$:

$$SxS = Sx.$$

We may deduce from this (since $1 \in S$) that $xS \subseteq Sx$ for *all* x in G. At this point the customary substitution of x^{-1} for x gives $Sx \subseteq xS$, and it follows that S is a normal subgroup of G. □

DEFINITION. The *factor group* (or *quotient group*) of the group G with respect to its normal subgroup N is the group formed by the cosets of N in G with the operation of complex multiplication.

We denote this factor group by the symbol G/N. A factor group G/N is called *proper* if $N \neq G$ and $N \neq 1$.

Note that if G is any group then the order of G/N is the index of N in G, by the definitions. We stress again the nature of the factor group G/N of G by pointing out that G/N is a set of sets of elements of G, and *not* a subset of G.

Example 4.05. Let Z be the additive group of integers and let N_n be the subset consisting of all multiples of the integer n, which by the way we shall suppose to be greater than 1. It is easy to see that N_n is a subgroup, and since Z is abelian N_n is normal. The coset of N_n containing the integer a will be denoted (in additive notation) by N_n+a. It is, explicitly, the set $\{a+kn : k = 0, \pm 1,...\}$, or

$$\{..., a-n, a, a+n, a+2n,...\}.$$

This coset is a typical element of the factor group Z/N_n. A part of the definition of factor group is that the sum of N_n+a and N_n+b is $(N_n+a)+(N_n+b)$, which is, of course, $N_n+(a+b)$. This description is reminiscent of the group of residue classes modulo n, earlier denoted by Z_n. It is plain from the definitions that Z/N_n is indeed isomorphic to Z_n. This throws some light on the reasons why we attached importance to Z_n as a group example.

Example 4.06. What is the group G/N when $G = S_3$ and $N = \{1, (123), (132)\}$, in the usual notation? We proved earlier that N is a normal subgroup. The elements of G/N are N and Nx where $Nx = \{(12), (13), (23)\}$ and x is any one of the elements (12), (13), (23). The multiplication in G/N is easily described since $N^2 = N$;

$$N.N = N, \qquad N.Nx = Nx,$$
$$Nx.N = Nx, \qquad Nx.Nx = Nx^2 = N.$$

It follows that the group G/N has order 2. We mentioned this very example in Chapter 3, in connection with groups resulting from multiplication of complexes, though it could not then be called a factor group.

We have now reached a position in which we can relate homomorphisms and normal subgroups by means of factor groups. We remarked in the course of the proof of the previous theorem that if N is a normal subgroup of G then

$$NxNy = Nxy$$

for all x, y in G; in terms of homomorphisms, this is just the statement that the mapping ϕ, defined by $x\phi = Nx$, is a homomorphism from G onto G/N. The kernel is $\{x : Nx = N\}$, since N is the unit element of G/N; in other words, the kernel is N. This mapping is called the *natural homomorphism* from G onto G/N. The fact that a factor group defines a homomorphism has the following significant converse.

THEOREM 4.07. (First isomorphism theorem). *Let K be the kernel of the homomorphism ϕ from the group G onto the group H. Then H is isomorphic to G/K.*

Proof. We remark first that the factor group G/K exists since the kernel K is a normal subgroup of G, as shown in Theorem 4.03. Note that the homomorphism ϕ is required to be *onto* H.

We try to define a mapping ψ from G/K to H by specifying $x\phi$ as the image of the element Kx of G/K, where x is an arbitrary element of G:

$$(Kx)\psi = x\phi.$$

It is of course necessary to verify that $(Kx)\psi$ is independent of the choice of the representative x for the coset Kx. Suppose then that $Kx_0 = Kx$, that is, $x_0 = kx$ for some $k \in K$. The proposed definition would give

$$(Kx_0)\psi = x_0\phi.$$

But since k is in the kernel K of the homomorphism ϕ we have

$$x_0\phi = (kx)\phi = (k\phi)(x\phi) = x\phi,$$

$$(Kx)\psi = (Kx_0)\psi.$$

Thus the proposed definition of ψ is acceptable.

Next we show that ψ is a homomorphism. Take x, y in G. Then

$$(KxKy)\psi = (Kxy)\psi = (xy)\phi$$

since K is normal and $xK = Kx$. On the other hand

$$(Kx)\psi(Ky)\psi = (x\phi)(y\phi) = (xy)\phi$$

since ϕ is by hypothesis a homomorphism. Therefore

$$(KxKy)\psi = (Kx)\psi(Ky)\psi,$$

and ψ is a homomorphism from G/K into H.

We hope to show, of course, that ψ is an isomorphism from G/K onto H. To this end it is enough to prove that ψ is both one–one and onto, by Theorem 2.16. Suppose

$$(Kx)\psi = (Ky)\psi,$$

and let us deduce that $Kx = Ky$. We have at once $x\phi = y\phi$. Since ϕ is a homomorphism

$$(xy^{-1})\phi = (x\phi)(y\phi)^{-1} = 1,$$

which implies that $xy^{-1} \in K$. It follows that $Kx = Ky$, and so ψ is one–one. Finally we show that ψ is a mapping *onto* H. Since ϕ is known to be *onto* H we see that an arbitrary element of H has the form $x\phi$, for a suitable element x of G. Clearly Kx maps onto $x\phi$ under ψ. Hence ψ is onto H.

This proves that ψ is an isomorphism from G/K onto H, and the theorem is proved. □

COROLLARY 4.08. *The homomorphism ϕ from G to H is one–one if and only if $K = 1$.* □

This remarkable theorem allows us to determine all the homomorphic images H of G, up to isomorphism. A typical image H is found by taking a normal subgroup N of G and forming G/N, since we have displayed a one–one correspondence between the homomorphic images of G and the factor groups of G.

In the most general situation of this sort we have a homomorphism ϕ of G into (but not necessarily onto) H. The image $G\phi$ of G under ϕ is, of course, a subgroup of H. Suppose we regard G as given and we seek information about H. Then, on the basis of the results we already have, we can say that H has some subgroup which is isomorphic to some factor group of G.

Example 4.09. There is a homomorphism ϕ from the additive group Z of integers onto the additive group Z_n of residue classes modulo n; it is defined by $a\phi = [a]$ for all $a \in Z$. The kernel is $\{a : [a] = [0]\}$, that is, the set of all multiples of n. This is a normal subgroup N_n of Z, and we have already seen that $Z/N_n \cong Z_n$. The existence of this isomorphism is, of course, just what we would expect from the first isomorphism theorem.

The remaining ideas of this chapter will be arranged rather haphazardly around the ideas of normal subgroup and homomorphism. They are still of importance, but not of the same order of difficulty as some fundamental ideas yet to appear.

One natural line of inquiry is concerned with how far from being normal a subgroup can be. The subgroup N of the group G is normal if and only if $x^{-1}Nx$ takes only one value, namely N, as x varies in G. We define a 'normal complex' in a similar way. For a general complex we have the concept of conjugates, defined as follows.

Definition. A *conjugate* of the complex C in the group G is any complex of the form $x^{-1}Cx$ for some element x in G. We shall write $x^{-1}Cx$ as C^x.

For the most part we shall consider either one-element complexes or subgroups. In the former case we identify the complex $\{y\}$ with the group element y which it contains, and we write $x^{-1}yx$ as y^x. The number of conjugates of a given subgroup is a rough measure of how near to or far from being normal that subgroup is.

Example 4.10. Take S_3, as usual, and ask what the conjugates of the element (123) are. It will be seen that

$$(12)^{-1}(123)(12) = (132),$$

so (132) is certainly one such conjugate. Since conjugate elements (clearly) have the same order, and since S_3 has only two elements of order three, (123) has no further conjugates. The subgroup gp{(123)}, being normal, has only one conjugate, namely itself. It will be found, however, that the subgroups generated by the elements (12), (13), (23) respectively are conjugate in pairs. The sort of calculation needed to prove this is:

$$(123)^{-1}\{1, (12)\}(123) = \{1, (23)\}.$$

Suppose we define the relation $C \sim D$ between complexes C and D in a fixed group G to hold if and only if C is conjugate to D. The proof that this is an equivalence relation is surely easy enough for the reader to supply. The resulting equivalence classes are called *conjugacy classes*. Such classes of elements and subgroups may be investigated and exploited by means of normalizers.

DEFINITION. The *normalizer* of the (non-empty) complex C in the group G is the complex $\{x : x^{-1}Cx = C\}$ of G. This normalizer will be denoted† by $N(C)$.

We note that it is not necessary for C and $N(C)$ to commute element by element—only commutation of these as complexes is required. The normalizer of a one-element complex is often called the *centralizer* of that complex.

Example 4.11. The normalizer of the subgroup of S_3 generated by (123) is S_3 itself, since the subgroup is normal. Some calculation will show that the subgroup generated by (12) is its own normalizer. The centralizer of the *element* (123) can easily be shown to be $\{1, (123), (132)\}$, and the centralizer of (12) is $\{1, (12)\}$.

THEOREM 4.12. *The normalizer $N(C)$ of the (non-empty) complex C in the group G is a subgroup of G. If T is any right transversal of $N(C)$ in G then the distinct conjugates of C in G are* $\{t^{-1}Ct : t \in T\}$.

Proof. It is easy to prove this theorem by routine arguments, and we leave the reader to do so in detail. The proof that we

† In this book mappings are written on the right, as a rule. Consistency therefore suggests CN for the normalizer of C. However, in this and other instances, the break with tradition and habit seems too great to give consistency its due.

shall give here is more abstract; perhaps it will give additional insight.

We define an equivalence relation in G as follows. If $a, b \in G$ then $a \sim b$ means that

$$a^{-1}Ca = b^{-1}Cb.$$

The verification that this *is* an equivalence relation is almost trivial. The equivalence class $[1]$ is precisely $N(C)$; and if $a \sim b$ then $ax \sim bx$.

The required result will then follow from a theorem which we state separately, as follows. □

THEOREM 4.13. *Let an equivalence relation, $\sim$, be defined in a group G, and suppose that if $a \sim b$ then $ax \sim bx$ for all a, b, x in G. Then $[1]$ is a subgroup of G, and the equivalence classes are the right cosets of $[1]$ in G.*

Proof. We use Theorem 3.06 in showing that $[1]$ is a subgroup. Clearly $[1]$ is non-empty, as $1 \sim 1$ and $1 \in [1]$. Let $x, y \in [1]$; thus

$$x \sim 1, \qquad y \sim 1.$$

On multiplication by y^{-1}, the latter gives $1 \sim y^{-1}$ and so $y^{-1} \sim 1$. Then

$$xy^{-1} \sim 1y^{-1} \sim y^{-1} \sim 1,$$

that is $xy^{-1} \in [1]$. Therefore $[1]$ is a subgroup.

Because $a \sim b$ implies $ab^{-1} \sim 1$ the equivalence classes are precisely the right cosets of $[1]$, by definition of right coset. □

The fact that $N(C)$ is a subgroup allows us to say that $N(C)$ is the largest subgroup of G in which C is normal, provided C itself is a subgroup. Two consequences of Theorem 4.12 are worth recording.

COROLLARY 4.14. *The number of conjugates of C in G equals the index of $N(C)$ in G.* □

COROLLARY 4.15. *The number of conjugates of C in the finite group G divides the order of G.*

Proof. Use Corollary 4.14 and Lagrange's theorem 3.15. □

THEOREM 4.16. *If G is a group with a proper subgroup of finite index then G has a proper (or trivial) normal subgroup of finite index.*

Proof. Let S be a proper subgroup of G and consider its normalizer $N(S)$. Since $N(S) \geqslant S$, the index $|G:N(S)|$ divides $|G:S|$. At all events $|G:N(S)|$ is finite, and so S has only finitely many conjugates by Corollary 4.14.

Let N denote the intersection of all these conjugates. Then N is a subgroup of finite index in G, by Theorems 3.07 and 3.20. We now prove normality. Let $n \in N$ and $x \in G$. It is clear that n lies in every conjugate of S, and it follows that n^x has the same property. Therefore $n^x \in N$, and so $N^x \subseteq N$. This suffices to show that $N \trianglelefteq G$. Clearly N is a proper subgroup of G, like S. □

This theorem tells us nothing new about finite groups, of course; there we may always take $N = 1$. It is non-trivial for infinite groups.

Associated with each class of conjugate complexes in the group G there is a homomorphic image of G. Let $\{C_\lambda : \lambda \in \Lambda\}$ be a complete set of conjugates, Λ being a suitable index set. The arbitrary element x of G defines a mapping of the set $\{C_\lambda\}$ into itself, the image of C_λ being C_λ^x. This mapping is one–one and onto; for if $C_\lambda^x = C_\mu^x$ then $C_\lambda = C_\mu$, and clearly the complex $C_\lambda^{x^{-1}}$ maps onto C_λ. Thus each group element x determines a permutation of $\{C_\lambda\}$. We now define a mapping ϕ of G into the symmetric group on the set $\{C_\lambda\}$ by stating that $x\phi$ is to be the permutation corresponding to the element x of G. Then ϕ is a homomorphism, since

$$C_\lambda^{xy} = y^{-1}x^{-1}C_\lambda xy = y^{-1}C_\lambda^x y = (C_\lambda^x)^y.$$

The following characterization of normal complexes in terms of conjugate elements is often useful.

THEOREM 4.17. *The (non-empty) complex C of the group G is normal if and only if it is the union of complete conjugacy classes of elements of G.*

Proof. Suppose C is a non-empty complex and is also the union of certain conjugacy classes; we mean, of course, the set-theoretic union. This implies that if x lies in C then so does x^y for all y in G. Thus $y^{-1}Cy \subseteq C$, which suffices to show that C is normal.

On the other hand, it is clear that if x lies in C and C is normal then x^y lies in C for all y in G. Thus C contains all conjugates of each of its elements, and is the union of complete conjugacy classes. □

COROLLARY 4.18. *The subgroup N of the group G is normal if and only if it has a set of generators which is the union of complete conjugacy classes of elements of G.*

Proof. Suppose N is a subgroup with a set of generators of the kind specified. An arbitrary (non-identity) element x of N may be represented as $x_1^{\epsilon_1} \dots x_n^{\epsilon_n}$ where $x_1, \dots, x_n$ are some of these generators and $\epsilon_i = \pm 1$ for $1 \leqslant i \leqslant n$; this is possible by Theorem 3.10. Note that

$$(ab)^x = x^{-1}abx = x^{-1}ax.x^{-1}bx = a^x b^x,$$

$$(a^{-1})^x = x^{-1}a^{-1}x = (x^{-1}ax)^{-1} = (a^x)^{-1};$$

these facts may be extended by induction to establish that

$$(x_1^{\epsilon_1} \dots x_n^{\epsilon_n})^y = (x_1^y)^{\epsilon_1} \dots (x_n^y)^{\epsilon_n}.$$

But (by hypothesis) x_i^y lies in the given set of generators for N, whatever element of G is chosen for y. Hence $y^{-1}xy \in N$, and Theorem 4.17 shows that N is normal in G.

The converse statement, which is almost trivial, may be dismissed with the remark that if N is a normal subgroup of G then the set of all elements of N is, by the theorem, a set of generators of the type sought. □

We turn from conjugate elements to conjugate subgroups.

THEOREM 4.19. *If S is any subgroup and x is any element of the group G then S^x is a subgroup isomorphic to S.*

Proof. We are not excluding the case in which S and S^x are identical; this will happen precisely when x is in $N(S)$. In general, let y, z be elements of S, so that $yz^{-1} \in S$. Then general elements of S^x will be $x^{-1}yx$, $x^{-1}zx$, and we have

$$(x^{-1}yx)(x^{-1}zx)^{-1} = x^{-1}(yz^{-1})x \in S^x$$

since $yz^{-1} \in S$. The subgroup criterion of Theorem 3.06 now shows that S^x is a subgroup, as clearly 1 lies in S^x.

We now set up an isomorphism from S onto S^x by defining $y\phi$ to be y^x for all y in S. It is almost trivial to verify that

$$(yz)\phi = (yz)^x = y^x z^x = (y\phi)(z\phi),$$

so that ϕ is a homomorphism. Further ϕ is one–one as $y^x = z^x$ implies $y = z$, for all y, z in S. And ϕ is onto S^x as y in S maps onto the general element y^x in S^x. Therefore, by Theorem 2.16, ϕ is an isomorphism as required. □

If he so desires the reader may construct many illustrations of the above theorems in S_3 and in other groups.

We next consider some special normal subgroups which, though present in every group, are sometimes trivial.

Definition. The *commutator* of the ordered pair of elements x, y in the group G is the element $x^{-1}y^{-1}xy$. It will be denoted by $[x, y]$.

This concept arises from asking how near the elements x, y come to commuting. We have the equation

$$xy = yxc$$

where $c = [x, y]$, and clearly $[x, y] = 1$ if and only if $xy = yx$; so a group is abelian if and only if every commutator in it equals 1.

Example 4.20. Which elements of S_3 are commutators? It is clear that every commutator, being of the form $x^{-1}y^{-1}xy$, must be the product of an *even* number of transpositions, and so only 1, (123), (132) could possibly occur as commutators. In fact each does occur:

$$[(12), (12)] = (12)(12)(12)(12) = 1,$$
$$[(12), (23)] = (12)(23)(12)(23) = (123),$$
$$[(12), (13)] = (12)(13)(12)(13) = (132).$$

Note that $x^{-1}y^{-1}xy$ may be written in either of the forms $x^{-1}x^y$, $(y^x)^{-1}y$. Therefore if either x or y lies in some normal subgroup of G then so does $[x, y]$.

Suppose next that the group G has normal subgroups N_1, N_2 with $N_1 \cap N_2 = 1$. Then the concept of commutator allows us to see easily that any element x of N_1 commutes with any element y of N_2, for $[x, y]$ lies in both N_1 and N_2 and is therefore 1.

Another interesting fact may be mentioned at this point: if $x^2 = 1$ for every element x of the group G then G is abelian. To prove this, take the commutator $[x, y]$ of elements x, y in G. Using in succession the facts that $x^2 = 1$, $y^2 = 1$, $(xy)^2 = 1$, we have

$$[x, y] = x^{-1}y^{-1}xy = xyxy = 1.$$

Since $xy = yx$ for each pair x, y of its elements, G is abelian.

Definition. The *derived group* (or *commutator subgroup*) of G is the subgroup generated by all the commutators in G. It will be denoted by $\delta(G)$; the notation G' is also current.

We emphasize that the derived group is defined to be a *subgroup* of G, not the *subset* of commutators in G. It is true that in S_3 and in other familiar groups the subgroup coincides with the subset. However, one must not suppose that these always coincide. (See Problem 5 following this chapter.)

The derived group of S_3 is $\text{gp}\{(123)\}$. The derived group is trivial if and only if the group in question is abelian. Now $\delta(G)$, or rather the order of $\delta(G)$, represents a measure of how near the group G comes to being abelian—the smaller $\delta(G)$ is, the closer G may be regarded to being abelian. In this sense a group G is as far removed as possible from being abelian if and only if $\delta(G) = G$, and we shall see that there do exist non-trivial groups with this property (see, for example, Problem 10 following this chapter; or Example 6.22; or Theorem 11.08).

Theorem 4.21. *Let N be a normal subgroup of the group G. Then G/N is abelian if and only if N contains $\delta(G)$.*

Proof. G/N is abelian if and only if $NxNy = NyNx$ for arbitrary elements x, y of G (so that Nx, Ny are arbitrary elements of G/N). This condition is equivalent to

$$Nxy = Nyx,$$

since N is a normal subgroup; and this holds if and only if $N = Nyxy^{-1}x^{-1}$. By Corollary 3.13 this is equivalent to $yxy^{-1}x^{-1} \in N$. Now N contains the arbitrary commutator $yxy^{-1}x^{-1}$ if and only if N contains all commutators; N being a subgroup, this is the case if and only if N contains the subgroup $\delta(G)$ generated by all the commutators in G. □

COROLLARY 4.22. *$G/\delta(G)$ is abelian.*

Proof. The corollary follows from the theorem as soon as we have proved the essential fact that $\delta(G)$ is a normal subgroup of G. It will suffice to prove that every conjugate of a commutator is a commutator and to apply Corollary 4.18 about a subgroup generated by complete classes of conjugate elements. But

$$[x, y]^z = [x^z, y^z]$$

as the left-hand side is $z^{-1}x^{-1}y^{-1}xyz$, and the right-hand side is $(z^{-1}xz)^{-1}(z^{-1}yz)^{-1}(z^{-1}xz)(z^{-1}yz)$. □

DEFINITION. The *centre* of the group G is the subset of those elements which commute with every element of G. It will be denoted by $\zeta(G)$; the notation $Z(G)$ is also current.

The centre could equally well be defined as the intersection of all the centralizers of the elements in G; or as the set of elements with no more than one conjugate each.

THEOREM 4.23. *Every subgroup S of G contained in the centre of G is a normal, abelian subgroup of G.*

Proof. Since each element of G commutes with each element of $\zeta(G)$ and so with each element of S, it follows that S is normal. The abelian property of S is obvious. □

COROLLARY 4.24. *The centre $\zeta(G)$ is a normal, abelian subgroup of G.*

Proof. It is necessary to prove only that $\zeta(G)$ is a subgroup. Clearly $1 \in \zeta(G)$. Let x, y lie in $\zeta(G)$ so that

$$xg = gx, \quad yg = gy$$

for all g in G. Then

$$(xy)g = xgy = g(xy).$$

Therefore $xy \in \zeta(G)$. The proof will be complete if we show that $x^{-1} \in \zeta(G)$. The equation $xg = gx$ implies

$$xgx^{-1} = g, \qquad gx^{-1} = x^{-1}g,$$

for all g in G, as required. □

The group G is abelian if and only if $G = \zeta(G)$, clearly. The size of $G/\zeta(G)$ therefore gives us another measure of the non-abelianness of G. The antithesis of an abelian group would be a group G for which $\zeta(G) = 1$, that is a group with trivial centre.

Example 4.25. The group S_3 has trivial centre. This may be proved by elementary (and by now familiar) calculations.

Example 4.26. What is the centre of the (multiplicative) group $\mathscr{M}_2$ of non-singular 2×2 matrices with real entries? Let $A \in \zeta(\mathscr{M}_2)$ where

$$A = \begin{pmatrix} a & b \\ c & d \end{pmatrix}.$$

Since A commutes with every element of $\mathscr{M}_2$ we have in particular that

$$\begin{pmatrix} a & b \\ c & d \end{pmatrix}\begin{pmatrix} -1 & 0 \\ 0 & 1 \end{pmatrix} = \begin{pmatrix} -1 & 0 \\ 0 & 1 \end{pmatrix}\begin{pmatrix} a & b \\ c & d \end{pmatrix}.$$

Some calculation gives $b = c = 0$. We have, further,

$$\begin{pmatrix} a & 0 \\ 0 & d \end{pmatrix}\begin{pmatrix} 0 & 1 \\ 1 & 0 \end{pmatrix} = \begin{pmatrix} 0 & 1 \\ 1 & 0 \end{pmatrix}\begin{pmatrix} a & 0 \\ 0 & d \end{pmatrix}.$$

It follows that $a = d$. Therefore $A = aI$ where I denotes the unit matrix. Since every aI with $a \neq 0$ does in fact lie in $\zeta(\mathscr{M}_2)$, this centre consists of all the multiples of I by real non-zero numbers.

THEOREM 4.27. *G is abelian if and only if $G/\zeta(G)$ is cyclic.*

Proof. The statement that if G is abelian then $G/\zeta(G)$ is cyclic hardly calls for proof. To prove the required converse, suppose that $a\zeta(G)$ is a generator of the cyclic group $G/\zeta(G)$, so that an arbitrary element of $G/\zeta(G)$ is $a^n\zeta(G)$, where n is some integer. Thus an arbitrary element of G has the form $a^n z$ where $z \in \zeta(G)$. Take another such element $a^m y$, with m a suitable integer and $y \in \zeta(G)$; then

$$a^m y . a^n z = a^m a^n yz = a^n a^m zy = a^n z . a^m y.$$

Since $a^m y$ and $a^n z$ commute, G is abelian. □

LEMMA 4.28. *If G is a finite group in which the index of every proper subgroup and of the trivial subgroup is divisible by the prime p, then $\zeta(G)$ has order divisible by p.*

Proof. If x is any element of G then its centralizer $C(x)$ is either a proper subgroup or, when $x \in \zeta(G)$, is G itself. By hypothesis and Corollary 4.14, therefore, either x has pk conjugates (for some positive k depending on x) or x has 1 conjugate. Suppose that $\zeta(G)$ has order n, and enumerate the elements of G by conjugacy classes. The conclusion is that G has $pk'+n$ elements for some $k' \geqslant 0$. On the other hand $|G|$ is divisible by p, by hypothesis. Therefore

$$pk'+n = |G|$$

shows that p divides n, which is what we set out to prove. □

THEOREM 4.29. *If G is a (finite) group of order p^n (where p is prime and n is a positive integer) then $\zeta(G) \neq 1$.*

Proof. This follows from Lagrange's Theorem 3.15 and Lemma 4.28. □

The foregoing theorem has a fundamental position in the theory of groups of p-power order. We shall apply it when $n = 2$:

COROLLARY 4.30. *Every group of order p^2 (where p is prime) is abelian.*

Proof. Let G have order p^2, so that $\zeta(G)$ has order 1 or p or p^2 by Lagrange's theorem. By the preceding theorem $\zeta(G)$

cannot have order 1. Suppose that $\zeta(G)$ has order p. Then $G/\zeta(G)$ has order p and is therefore cyclic; we have seen in Theorem 4.27 that if $G/\zeta(G)$ is cyclic then G is abelian, so $G = \zeta(G)$ and $\zeta(G)$ has order p^2. This contradicts the assumption that $\zeta(G)$ had order p. We are forced to the conclusion that $\zeta(G)$ has order p^2, which implies that G is abelian. □

The reader may care to search for a non-abelian group of order 2^3.

We now hint at some sort of 'duality' between $\delta(G)$ and $G/\zeta(G)$, based on the question of 'how far' G is from being abelian. Though $\delta(G) = 1$ if and only if $G = \zeta(G)$, it is not true that $\zeta(G) = 1$ implies $G = \delta(G)$ nor is it true that $G = \delta(G)$ implies $\zeta(G) = 1$. The group S_3 has $\zeta(S_3) = 1$ but $S_3 > \delta(S_3)$, as we have seen; an example to substantiate the other assertion will appear in Problem 9 of Chapter 11. The next theorem says something about the centre of a group G in which $G = \delta(G)$.

THEOREM 4.31. *If G is a group in which $G = \delta(G)$ then $G/\zeta(G)$ is a group with trivial centre.*

Proof. Let the centre of $G/\zeta(G)$ be $\zeta_2(G)/\zeta(G)$ and let a be any element of $\zeta_2(G)$. What we have to show is that $a \in \zeta(G)$, for then $G/\zeta(G)$ has centre $\zeta(G)/\zeta(G)$, that is $G/\zeta(G)$ has trivial centre.

We shall consider $[x, y]^a$ where x, y are arbitrary elements of G. Since $a\zeta(G)$ is central in $G/\zeta(G)$, we have $[g\zeta(G), a\zeta(G)] = \zeta(G)$ for any $g \in G$, which is equivalent to $[g, a]\zeta(G) = \zeta(G)$ and to $[g, a] \in \zeta(G)$. So there are elements z_1, z_2 in $\zeta(G)$ for which

$$x^{-1}x^a = z_1, \qquad y^{-1}y^a = z_2.$$

It follows that

$$\begin{aligned}
[x, y]^a &= (x^{-1}y^{-1}xy)^a \\
&= (x^a)^{-1}(y^a)^{-1}(x^a)(y^a) \\
&= (xz_1)^{-1}(yz_2)^{-1}(xz_1)(yz_2) \\
&= x^{-1}y^{-1}xy \\
&= [x, y].
\end{aligned}$$

Note the use of the fact that z_1 and z_2 are central.

We now see that a commutes with the commutator of any two elements x, y in G. Therefore a commutes with every element of the subgroup $\delta(G)$ generated by the commutators. But since $G = \delta(G)$ it follows that $a \in \zeta(G)$. □

Our next topic is centred around the concept of automorphism. That this has a place in the context of normal properties is shown by the following result.

THEOREM 4.32. *Let G be a group and let x be a fixed element of G. Then the mapping ϕ, defined by*

$$g\phi = g^x$$

for all g in G, is an automorphism of G.

Proof. This theorem is closely related to Theorem 4.19. Indeed, if we take $S = G$ in that theorem, then our Theorem 4.32 follows almost completely; we need show only that $G^x = G$. This may be done by observing that $g^{x^{-1}} \in G$ and that $g^{x^{-1}}\phi = g$, showing that $G^x \supseteq G$ and therefore that $G^x = G$. □

DEFINITION. *An inner automorphism* of the group G is any automorphism arising from conjugation; that is, of the form described in Theorem 4.32. Any other automorphism of G is called *outer*.

Any automorphism of an abelian group is either trivial or outer, obviously; so the automorphism α described in Example 2.23 is outer, as long as the abelian group G has some element of order neither 1 nor 2.

The next result shows that non-abelian groups have non-trivial inner automorphisms.

THEOREM 4.33. *Let $I(G)$ be the set of inner automorphisms of the group G. Then $I(G)$ is a group and is isomorphic to $G/\zeta(G)$.*

Proof. Let ϕ_x, ϕ_y be arbitrary elements of $I(G)$, and let them result from conjugation by x, y respectively. If $g \in G$ then

$$g(\phi_x \phi_y) = (g\phi_x)\phi_y = (g^x)\phi_y = (g^x)^y = g^{xy} = g\phi_{xy};$$

and so $\phi_x \phi_y = \phi_{xy}$. When $y = x^{-1}$ we have

$$\phi_x \phi_{x^{-1}} = \phi_1,$$

where ϕ_1 is the identity mapping on G, so $\phi_x^{-1} = \phi_{x^{-1}}$. Then

$$\phi_x \phi_y^{-1} = \phi_x \phi_{y^{-1}} = \phi_{xy^{-1}} \in I(G),$$

and Theorem 3.06 shows that $I(G)$ is a group.

Let ψ be the mapping G into $I(G)$ defined by

$$g\psi = \phi_g.$$

The previous paragraph shows that ψ is a homomorphism which is onto $I(G)$. The kernel of ψ is the set of elements that commute with every element of G—the centre of G. The first isomorphism theorem then shows that $G/\zeta(G) \cong I(G)$. □

DEFINITION. A *characteristic subgroup* of the group G is a subgroup that is mapped onto itself by every automorphism of G.

This means that the subgroup *as a whole* is left fixed; in general automorphisms will permute the individual elements of the subgroup.

It is an immediate consequence of the definitions that a characteristic subgroup is normal. Indeed, the normal subgroups of G are just those left fixed by $I(G)$. Not every normal subgroup is characteristic; we shall give an example to substantiate this assertion below.

THEOREM 4.34. *In any group G the subgroups $\delta(G)$ and $\zeta(G)$ are characteristic.*

Proof. Let α be an automorphism of G. Because α and α^{-1} are homomorphisms they each map the set of commutators into itself:

$$\begin{aligned}[x, y]\alpha = (x^{-1}y^{-1}xy)\alpha &= (x^{-1}\alpha)(y^{-1}\alpha)(x\alpha)(y\alpha) \\ &= (x\alpha)^{-1}(y\alpha)^{-1}(x\alpha)(y\alpha) = [x\alpha, y\alpha],\end{aligned}$$

and similarly for α^{-1}. It follows that α maps $\delta(G)$ onto $\delta(G)$; that is to say, $\delta(G)$ is characteristic.

Next let $z \in \zeta(G)$, so that $az = za$ for each a in G. If follows that $(a\alpha)(z\alpha) = (z\alpha)(a\alpha)$. Now $a\alpha$ is as arbitrary an element of G as a was, because α is onto G. Therefore $z\alpha \in \zeta(G)$. We thus have $\{\zeta(G)\}\alpha \subseteq \zeta(G)$, and the reverse inclusion follows from the fact that α^{-1} is an automorphism. Therefore $\zeta(G)$ is left fixed by α, and is a characteristic subgroup. □

THEOREM 4.35. *In any group G, a characteristic subgroup of a characteristic subgroup is characteristic in G, and a characteristic subgroup of a normal subgroup is normal in G.*

Proof. Let N be a subgroup of G, and suppose that N is left fixed by some set A of automorphisms of G. Consider the action of A on some characteristic subgroup K of N. If $\alpha \in A$ then α, on restriction to N, defines an automorphism of N, which we denote by α also. The characteristic property of K then ensures that $K\alpha = K$.

Now take N to be a characteristic subgroup of G, and A to be the full automorphism group of G. Since $K\alpha = K$ for all α in A, K is characteristic in G.

Secondly take N to be a normal subgroup of G, and A to be the group $I(G)$ of inner automorphisms of G. Again our earlier remarks give the required fact: $K\alpha = K$ for every $\alpha \in I(G)$, and so $K \trianglelefteq G$. □

The final concept to be studied in this chapter is more closely related to normal subgroups than may appear at first sight.

DEFINITION. The *direct product* of the finite set $\{G_i : 1 \leqslant i \leqslant n\}$ of groups is the set

$$\{(g_1,\ldots,g_n) : g_i \in G_i \text{ for } 1 \leqslant i \leqslant n\},$$

where $(g_1,\ldots,g_n)$ is an ordered set of group elements. Multiplication in the direct product is defined by

$$(g_1,\ldots,g_n)(h_1,\ldots,h_n) = (g_1 h_1,\ldots,g_n h_n)$$

where g_i, h_i lie in G_i. This direct product is denoted by $G_1 \times \ldots \times G_n$ or by $\prod_{1 \leqslant i \leqslant n}^{D} G_i$.

Theorem 4.36. *The direct product $G_1 \times \ldots \times G_n$ of the groups G_i for $1 \leqslant i \leqslant n$ is a group.*

Proof. We verify briefly the group axioms one by one. Closure follows immediately from the definition of direct product. The associative law is a consequence of the same law in each factor G_i. The unit element of the direct product is $(1, \ldots, 1)$, and the inverse of $(g_1, \ldots, g_n)$ is $(g_1^{-1}, \ldots, g_n^{-1})$; here, of course, the symbol 1 in the ith place denotes the identity element of G_i, and g_i^{-1} is the inverse of g_i in G_i. □

Example 4.37. Let C_2 and C_3 denote the groups of orders 2 and 3 respectively. Suppose a_m generates C_m, where $m = 2, 3$. Then the elements of $C_2 \times C_3$ are just the set of pairs

$$\{(a_2^i, a_3^j) : 1 \leqslant i \leqslant 2,\ 1 \leqslant j \leqslant 3\}.$$

Thus $C_2 \times C_3$ is a group of order 6. Now consider powers of the element of $C_2 \times C_3$ in which $i = j = 1$:

$$(a_2, a_3)^2 = (a_2^2, a_3^2) = (1, a_3^2),$$

$$(a_2, a_3)^3 = (a_2^3, a_3^3) = (a_2, 1),$$

$$(a_2, a_3)^4 = (a_2^4, a_3^4) = (1, a_3),$$

$$(a_2, a_3)^5 = (a_2^5, a_3^5) = (a_2, a_3^2),$$

$$(a_2, a_3)^6 = (a_2^6, a_3^6) = (1, 1).$$

Therefore (a_2, a_3) has six distinct powers. It follows that $C_2 \times C_3$ is cyclic; it is, or at least it is isomorphic to, the cyclic group of order six.† □

We remark that every group G is isomorphic to the direct product of itself and the group of order 1; we call such a direct product trivial, and usually interest ourselves only in non-trivial

† The learner who is now struggling may find some comfort in the story that Cayley supposed that C_6 and $C_2 \times C_3$ were non-isomorphic.

direct products. Note that if the groups G_i are finite, of order γ_i for $1 \leqslant i \leqslant n$, then $\prod^D_{1 \leqslant i \leqslant n} G_i$ has order $\prod_{1 \leqslant i \leqslant n} \gamma_i$; this is implied by the very definition of direct product.

THEOREM 4.38. *The direct product G of the set $\{G_i : 1 \leqslant i \leqslant n\}$ of groups contains subgroups H_i which are isomorphic to G_i, for $1 \leqslant i \leqslant n$, and which have the following properties:*

(i) *H_i commutes with H_j element by element if $1 \leqslant i < j \leqslant n$;*

(ii) *$G = H_1 \ldots H_n$;*

(iii) *$H_i \cap H_1 \ldots H_{i-1} H_{i+1} \ldots H_n = 1$ for $1 \leqslant i \leqslant n$.*

Proof. We define H_i to be the subset of elements of G with the property that $g_j = 1$ if $1 \leqslant j \leqslant n$ and $j \neq i$, the element g_i taking an arbitrary value in G_i. The facts that this is a subgroup and that it is isomorphic to G_i scarcely call for proof.

(i) It is clear that $(1,\ldots, 1, g_i, 1,\ldots, 1)$ and $(1,\ldots, 1, g_j, 1,\ldots, 1)$ commute provided $i < j$; the product is

$$(1,\ldots, 1, g_i, 1,\ldots, 1, g_j, 1,\ldots, 1),$$

in whichever order the factors are taken.

(ii) Trivially, $H_1 \ldots H_n$ is the complex product of $H_1,\ldots, H_n$. We are required to show that the arbitrary element $(g_1,\ldots, g_n)$ of G can be expressed in the form $h_1 \ldots h_n$ with $h_i \in H_i$ for $1 \leqslant i \leqslant n$. We choose h_i to be $(1,\ldots, 1, g_i, 1,\ldots, 1)$, and the rest is easy.

(iii) The elements of $H_1 \ldots H_{i-1} H_{i+1} \ldots H_n$ are just those elements of G with ith component equal to 1, and the only such element in H_i is $(1,\ldots, 1)$, which is the identity element of G. □

Thus far we have examined a procedure which constructs a group G from *given* groups $G_1,\ldots, G_n$. We now suppose that we have been given a group G instead and that we are seeking information about whether it is, or (strictly) is isomorphic to, some direct product.

THEOREM 4.39. *Let the group G have subgroups G_i for $1 \leqslant i \leqslant n$ satisfying the conditions:*

(i) *G_i commutes with G_j element by element if $1 \leqslant i < j \leqslant n$;*

(ii) $G = G_1 \ldots G_n$;

(iii) $G_i \cap G_1 \ldots G_{i-1} G_{i+1} \ldots G_n = 1$ *for* $1 \leqslant i \leqslant n$.

Then $G \cong G_1 \times \ldots \times G_n$.

Proof. The arbitrary element g of G can, by (ii), be written as $g_1 \ldots g_n$, where $g_i \in G_i$ for $1 \leqslant i \leqslant n$. The hypotheses imply that this representation is unique. Suppose that we had

$$g = g_1 \ldots g_n = g'_1 \ldots g'_n$$

where, of course, $g'_i \in G_i$. We may obtain the equation

$$g_i(g'_i)^{-1} = g_1^{-1} g'_1 \ldots g_{i-1}^{-1} g'_{i-1} g_{i+1}^{-1} g'_{i+1} \ldots g_n^{-1} g'_n$$

by means of (i). But then (iii) shows that $g_i(g'_i)^{-1} = 1$, that is $g_i = g'_i$, and this holds for $1 \leqslant i \leqslant n$. Hence the representation of g is unique.

We now construct a mapping ϕ from G to $G_1 \times \ldots \times G_n$ by putting

$$g\phi = (g_1, \ldots, g_n).$$

If $h = h_1 \ldots h_n$ where $h \in G$ and $h_i \in G_i$ for $1 \leqslant i \leqslant n$, then

$$gh = g_1 \ldots g_n h_1 \ldots h_n = (g_1 h_1) \ldots (g_n h_n)$$

by (i), and so

$$\begin{aligned}(gh)\phi &= (g_1 h_1, \ldots, g_n h_n)\\ &= (g_1, \ldots, g_n)(h_1, \ldots, h_n)\\ &= (g\phi)(h\phi);\end{aligned}$$

here we use the definition of group multiplication in $G_1 \times \ldots \times G_n$. Therefore ϕ is a homomorphism.

But ϕ is one–one and onto. It is one–one because if $g\phi = h\phi$ then

$$(g_1, \ldots, g_n) = (h_1, \ldots, h_n),$$

in the above notation, $g_i = h_i$ for $1 \leqslant i \leqslant n$, and so $g = h$. It is onto because $g_1 \ldots g_n$ maps onto $(g_1, \ldots, g_n)$. Therefore ϕ is an isomorphism from G onto $G_1 \times \ldots \times G_n$, and the theorem is now proved. □

Example 4.40. Consider the set of all vectors in the plane $\mathscr{P}$, or, what comes to the same thing, the set of all 1×2 matrices with real entries (see Examples 1.18 and 1.32); the group operation is addition. An easy application of the above theorem shows that this group is isomorphic to $R_1 \times R_2$ where R_1, R_2 are each isomorphic to the additive group of real numbers. Specifically, R_1 is $\{(r, 0): r \text{ real}\}$, and R_2 is $\{(0, r): r \text{ real}\}$.

Example 4.41. Let L be a fixed line in the plane $\mathscr{P}$, and consider the group $\mathscr{G}$ of all congruence mappings of $\mathscr{P}$ which leave L fixed. (Elements of $\mathscr{G}$ need not fix L point by point, but they must map L onto L.) It is geometrically clear that $\mathscr{G}$ is generated by a subgroup $\mathscr{D}$ of the translations of $\mathscr{P}$ and by the group $\mathscr{R}$ generated by the reflection in L; indeed $\mathscr{G} = \mathscr{D}\mathscr{R}$. We have further that $\mathscr{G} \simeq \mathscr{D} \times \mathscr{R}$ because $\mathscr{G}$ is abelian, so condition (i) of the theorem holds, (ii) has just been mentioned, and (iii) is obvious.

Example 4.42. Let $\mathscr{M}$ be the set of all $(n+1)\times(n+1)$ matrices with real entries and of the form $I + \sum_{i=1}^{n} a_i E_{i,n+1}$, where I is the unit $(n+1)\times(n+1)$ matrix and $E_{i,j}$ is the matrix with every entry 0 except for 1 in row i, column j. It will be found that

$$\Big(I + \sum_{i=1}^{n} a_i E_{i,n+1}\Big)\Big(I + \sum_{i=1}^{n} b_i E_{i,n+1}\Big) = I + \sum_{i=1}^{n} (a_i + b_i) E_{i,n+1}.$$

The reader will easily be able to show that $\mathscr{M}$ is an abelian group under matrix multiplication, and that in view of the above equation

$$\mathscr{M} \simeq \mathscr{N}_1 \times \ldots \times \mathscr{N}_n$$

where $\mathscr{N}_i$ is the set of all matrices of the form $I + a_i E_{i,n+1}$. In fact $\mathscr{M}$ is closely related to the group of translations in $\mathscr{P}$.

As an example of a group which is not a (non-trivial) direct product we mention the cyclic group of order 4. It has only one proper subgroup, and so cannot be expressed as any product of proper subgroups.

There is an alternative form of Theorem 4.39 in which condition (i) is replaced by:

(i′) $G_i \trianglelefteq G$ for $1 \leqslant i \leqslant n$.

The reader will find it an easy matter to verify that this new version is equivalent to that originally given.

We stress again that the data in Theorems 4.38 and 4.39 are

completely different, however similar the theorems are in form. In the former case we are given groups $G_1, \ldots, G_n$ and then consider properties of their direct product, while in the latter we start with a group G and postulate subgroups of it with certain interesting properties. Sometimes G is called an *external* direct product or an *internal* direct product, as the case may be.

Example 4.43. We close this chapter with an example of a group containing a normal subgroup that is not characteristic. Take isomorphic non-trivial groups G_1, G_2 and form their direct product G. By Theorem 4.38, G contains subgroups H_1, H_2 isomorphic to G_1. It is clear that H_1 is normal, but it is not characteristic because an automorphism of G can be constructed to map H_1 onto H_2 and H_2 onto H_1. Details are left for the reader to supply.

Problems

1. Prove that the quaternion group Q_8 (Example 1.30) has only one element of order 2. Determine the centre of Q_8, and prove that all subgroups of Q_8 are normal. Does Q_8 have any factor group isomorphic to the cyclic group of order 4?

2. Find all classes of conjugate elements and hence all normal subgroups in the alternating group A_4 on $\{1, 2, 3, 4\}$. What is the centre of this group? What is its derived group?

3. Prove that, in the symmetric group S_4 on the set $\{1, 2, 3, 4\}$, the subset

$$V = \{1,\ (12)(34),\ (13)(24),\ (14)(32)\}$$

forms a normal subgroup of order 4. Show, by a suitable choice of transversal of V in S_4 or otherwise, that S_4/V is isomorphic to the symmetric group S_3 on $\{1, 2, 3\}$.

4. Which of the following multiplicative groups of residues are isomorphic to (non-trivial) direct products?

   (i) $\{[1], [3], [5], [7]\}$ modulo 8;

   (ii) $\{[1], [3], [9], [11]\}$ modulo 16;

   (iii) $\{[1], [7], [9], [15]\}$ modulo 16.

*5. Let G be the set

$$\{(a_1, a_2, a_3, a_4) : a_1, a_2, a_3, a_4 \text{ are integers modulo } p^2\}$$

where p is any prime. The product of the elements (a_1, a_2, a_3, a_4) and (b_1, b_2, b_3, b_4), taken in that order, is defined to be

$$(a_1+b_1+pa_3b_1,\ a_2+b_2+pa_4b_2,\ a_3+b_3+pa_3b_2,\ a_4+b_4+pa_2b_1).$$

(i) Prove that G is a group under this multiplication.

(ii) Find the form of the commutator of two general elements in G, and deduce that $\delta(G) \leqslant \zeta(G)$.

(iii) Show that both $(p, 0, 0, 0)$ and $(0, p, 0, 0)$ are commutators whereas their product is not.

6. The group G has a subgroup S with the property that every right coset of S in G is also a left coset of S in G. Must S be a normal subgroup of G?

7. Let S denote the symmetric group on the set of integers. Show that those permutations which each move only a finite number of elements form a normal subgroup of S.

8. (i) Let $\mathscr{D} = \text{gp}\{A, B\}$ where

$$A = \begin{pmatrix} 1 & 0 \\ 0 & -1 \end{pmatrix}, \qquad B = \begin{pmatrix} 1 & 1 \\ 0 & -1 \end{pmatrix}$$

as in Problem 5 of Chapter 3. Is $\text{gp}\{AB\}$ normal in $\mathscr{D}$?

(ii) Let $\mathscr{G} = \text{gp}\{A, B\}$ where

$$A = \begin{pmatrix} 1 & 1 \\ 0 & 1 \end{pmatrix}, \qquad B = \begin{pmatrix} 2 & 0 \\ 0 & 1 \end{pmatrix}$$

as in Problem 11 of Chapter 3. Is $\text{gp}\{A\}$ normal in $\mathscr{G}$?

9. The centralizer $C(X)$ of the complex X in the group G is defined to be $\{y : y \in G,\ [x, y] = 1 \text{ for } all\ x \text{ in } X\}$. Prove that

$$C(X) = \bigcap_{x \in X} C(x)$$

where $C(x)$ denotes the centralizer of $\{x\}$ in G.

Prove the following statements about the centralizers of the complexes X, Y, S in G:

(i) If $X \subseteq Y$ then $C(X) \supseteq C(Y)$.

(ii) $S \subseteq C\{C(S)\}$.

(iii) $C(S) = C[C\{C(S)\}]$.

10. Prove that every element of A_5 is a commutator.

11. Prove that the multiplicative group of the non-zero complex numbers is isomorphic to the direct product of the multiplicative group of positive real numbers and the additive group of real numbers modulo 2π (see Problem 10 of Chapter 2).

12. (i) Let G be the subgroup of S_6 generated by the set

$$\{(12)(45),\ (12)(46),\ (13)(45),\ (13)(46)\}.$$

Prove that G has a normal subgroup of order 9, and find the index of this in G.

(ii) The dihedral group of order 12 is defined in Example 1.25 as a set of mappings leaving a 6-sided regular polygon invariant. Prove that the group is isomorphic to a non-trivial direct product.

13. Let $I(G), A(G)$ be the groups of inner, all automorphisms respectively of the group G. Prove that $I(G)$ is normal in $A(G)$.

14. Prove that in any group G the set

$$\{x : x \text{ has only finitely many conjugates in } G\}$$

forms a subgroup. Prove further that this subgroup is characteristic in G.

15. Let G be the direct product of G_1 and G_2, and let N_i be a normal subgroup of G_i for $i = 1, 2$. Which of the following statements are true?

(i) $\delta(G_1 \times G_2) \cong \delta(G_1) \times \delta(G_2)$ and $\zeta(G_1 \times G_2) \cong \zeta(G_1) \times \zeta(G_2)$.

(ii) $G_1 \times G_2$ has a normal subgroup isomorphic to $N_1 \times N_2$.

(iii) $(G_1 \times G_2)/(N_1 \times N_2) \cong G_1/N_1 \times G_2/N_2$.

(iv) If G_1, G_2 have no proper normal subgroups then any proper normal subgroup of $G_1 \times G_2$ must be isomorphic to G_1 or G_2.

(v) All normal subgroups of $G_1 \times G_2$ have the form $N_1 \times N_2$ where $N_i \trianglelefteq G_i$ for $i = 1, 2$.

***16.** (i) The normal subgroup N of the group G is such that $N \cap \delta(G) = 1$. Prove that $N \leqslant \zeta(G)$, and deduce that $\zeta(G/N) = \zeta(G)/N$.

(ii) Prove that if N_1, N_2 are normal subgroups of the group G then $G/(N_1 \cap N_2)$ is isomorphic to a subgroup of $G/N_1 \times G/N_2$.

****17.** Prove that, if $\{C_\lambda : \lambda \in \Lambda\}$ is a complete set of conjugate complexes in the group G, then the kernel K of the homomorphism which takes x in G onto the permutation replacing C_λ by C_λ^x is given by

$$K = \bigcap_{\lambda \in \Lambda} N(C_\lambda).$$

Show that $N(C_\lambda^x) = N(C_\lambda)^x$ for each λ, and deduce that K also equals $\bigcap_{x \in G} N(C)^x$ where C is a complex arbitrarily selected from the given conjugate set.

18. Find a complete class C of conjugate elements each of order 3 in the alternating group A_4 on $\{1, 2, 3, 4\}$. Consider the homomorphism (as described in Problem 17) from A_4 into the symmetric group on C, and find its kernel.

**19. Let S be a subgroup of the group G. Show that if g is a fixed element of G, then the mapping that takes the coset Sx to Sxg is a permutation of the right cosets of S in G. Hence define a mapping ϕ from G into the symmetric group on these cosets. Show that ϕ is a homomorphism and that its kernel is the intersection of all the conjugates of S in G.

*20. Let $N(X)$, $C(X)$ denote respectively the normalizer, centralizer of the (non-empty) complex X in the group G (see Problem 9 above). Prove that $C(X) \trianglelefteq N(X)$.

Prove further that if X is a subgroup of G then $N(X)/C(X)$ is isomorphic to a group of automorphisms of X.

Deduce Theorem 4.33.

5

Finitely generated abelian groups

AT the present day, to determine in any useful sense all finite groups is a problem beyond the resources of group theory, largely because the structure of these groups is so varied; the classification of all infinite groups is much harder still. To obtain structure theorems it is therefore necessary to look at rather restricted classes of groups. In Chapter 3 (Theorem 3.21 and Problem 12) we gave a description of all finite cyclic groups; in such groups every subgroup is cyclic, and every number compatible with Lagrange's theorem occurs as the order of a certain unique subgroup. We are now going to solve corresponding structure problems for all finite abelian groups. In fact we shall do more. It will appear that it is more natural to consider the finitely generated abelian groups, which of course include the finite abelian groups as a subset, because much the same tools and methods will determine their structure also. We shall take advantage of this fact. We should perhaps caution the reader that, in the *non-abelian* case, finitely generated groups are significantly more difficult to treat than finite groups.

Unfortunately, the notation traditionally used for abelian groups is not the multiplicative notation of the last two chapters. An additive notation is usual—the binary operation is called 'addition' instead of 'multiplication', and one writes $a+b$ for the 'sum' of the elements a, b in an abelian group instead of writing ab for their 'product'. This new notation is logical enough when one takes into account the existence of rings. In these objects there are two binary operations. A ring must form an abelian group with respect to one operation but need not have this property with respect to the other. In order to have the ring axioms, especially the distributive laws, resemble the familiar laws holding for the real numbers, it is necessary to use addition for the abelian group operation and multiplication for the other. We adhere to this practice in the study of abstract abelian groups for the sake of uniformity.

This additive notation has certain consequences of note. The identity element of the abelian group A is written 0 (and called 'zero'), while the inverse of a in A is written as $-a$. We write na instead of a^n, for arbitrary integers n. In correspondence with Theorem 2.07 we have $-(na) = n(-a)$ for such n, and we define $(-n)a$ as, say, $-(na)$; of course, we define $0a$ to be 0 for all a. The statement that $ma+na = (m+n)a$ for all integers m, n is the new version of Theorem 2.08 according to which $a^m a^n = a^{m+n}$. Note carefully that we have

$$n(a+b) = na+nb$$

by virtue of the fact that the elements a, b of A commute—the corresponding relation $(ab)^n = a^n b^n$ does not hold generally in non-abelian groups. Theorem 2.09 implies that $m(na) = (mn)a$.

The product $C_1 C_2$ of the complexes C_1, C_2 becomes their sum C_1+C_2. In place of the direct product of groups $G_1,\ldots, G_n$ we have their direct sum, written as $\sum_{i=1}^{n}{}^{D}\, G_i$ or $G_1 \oplus \ldots \oplus G_n$; note the need for a special notation for direct sum to distinguish it from sum. Sometimes the quotient group G/N is written as $G-N$ and called a difference group, but we shall persist with the term 'factor group' and the symbol G/N.

Our basic tools in the task at hand are cyclic groups and direct sums. Since every subgroup of an abelian group is normal, it is perhaps to be expected that direct sums should appear in view of the criterion in Theorem 4.39 for a group to be isomorphic to the direct sum of certain subgroups. We observe that the direct sum of any (finite) set of cyclic groups is an abelian group. Thus to find an abelian group of order p^n (where p is prime) we may take cyclic groups of orders $p^{n_1},\ldots, p^{n_r}$ where

$n_1,\ldots,n_r$ are any positive integers for which $n_1+\ldots+n_r = n$, and take their direct sum; and a similar construction may be carried out in more general cases. We know, of course, that cyclic groups of all orders exist.

The fundamental result on finitely generated abelian groups is a converse of this. It states that every such group is the direct sum of cyclic groups (finite or infinite), and most of this chapter is devoted to its proof. Many proofs of this result are known. The method that we have chosen is more constructive and may be easier for the learner to follow, as well as being more capable of generalization, than other briefer but less lucid proofs that have been found. (See Problems 15 and 16 following this chapter.)

The following definition applies not only to abelian groups but to general (even infinite) groups.

DEFINITION. A *p-group* (where p is any prime number) is a group in which the order of each element is some power of p.

By Corollary 3.16 to Lagrange's theorem any group of order p^n is a p-group, and we shall prove in the course of Chapter 7 that every finite p-group has p-power order. It will be of some help to prove the latter statement at once in the abelian case. Let A be a finite abelian p-group. If A has no proper subgroup then A is cyclic, A is generated by an element of p-power order, and so A has p-power order. This is the first step in an inductive proof on the order of A. If A has a proper subgroup B then both $|B|$ and $|A/B|$ are strictly less than $|A|$, which implies that B and A/B each have p-power order. Since $|A| = |B||A/B|$, it follows that A has p-power order.

We now tackle finite abelian groups.

THEOREM 5.01. *Let A be a finite abelian group in which $\{p_1,\ldots,p_n\}$ is the set of primes occurring as orders of elements. Then the elements of A of p_i-power order form a subgroup, A_i say; and* $A \simeq \sum_{i=1}^{n}{}^{D} A_i$.

Proof. We prove that A_i is a subgroup by applying the usual subgroup criterion. Let x, y be elements of A_i with orders p_i^r, p_i^s respectively, where r, s are certain non-negative integers; we must consider $x-y$. If t is the larger of r and s then

$$p_i^t(x-y) = p_i^t x - p_i^t y = p_i^{t-r}(p_i^r x) - p_i^{t-s}(p_i^s y) = 0$$

since $p_i^r x = p_i^s y = 0$. It follows from $p_i^t(x-y) = 0$ that the order of $x-y$ divides p_i^t, and so is certainly a power of p_i. Since A_i is non-empty by hypothesis, we see that A_i is a subgroup of A.

We prove that A is a direct sum of the required form by applying the criterion of Theorem 4.39 for a direct sum.

First, it is clear, because A is abelian, that each element of A_i commutes with each element of A_j whenever $i \neq j$.

Secondly, we have to show that $A = A_1 + \ldots + A_n$; that is, that an arbitrary element a of A has an expression of the form $a_1 + \ldots + a_n$ where $a_i \in A_i$ for each i. Let a have order m, and suppose that the prime q divides m. Then $a^{m/q}$, which has order q, still lies in A; and it follows that $q \in \{p_1, \ldots, p_n\}$. Therefore $m = p_1^{\alpha_1} \ldots p_n^{\alpha_n}$ for suitable non-negative integers $\alpha_1, \ldots, \alpha_n$. The Euclidean algorithm in its general form† asserts the existence of integers $\beta_1, \ldots, \beta_n$ for which

$$\beta_1 \frac{m}{p_1^{\alpha_1}} + \ldots + \beta_n \frac{m}{p_n^{\alpha_n}} = 1$$

because the integers $\dfrac{m}{p_1^{\alpha_1}}, \ldots, \dfrac{m}{p_n^{\alpha_n}}$ have 1 for their highest common factor. It follows that

$$a = \beta_1 \frac{m}{p_1^{\alpha_1}} a + \ldots + \beta_n \frac{m}{p_n^{\alpha_n}} a.$$

We may take a_i to be $\beta_i \dfrac{m}{p_i^{\alpha_i}} a$, for then

$$p_i^{\alpha_i} a_i = \beta_i(ma) = \beta_i 0 = 0,$$

and $a_i \in A_i$ as a_i thus has p_i-power order.

Thirdly, we must prove that, for each i, A_i has trivial intersection with $A_1 + \ldots + A_{i-1} + A_{i+1} + \ldots + A_n$. This follows from

† See the Appendix.

the obvious fact that 0 is the only element of p_i-power order in this sum, and so is the only element in both A_i and the sum.

Therefore $A \simeq \sum_{i=1}^{n}{}^{D} A_i$, as required. □

COROLLARY 5.02. *A finite abelian group contains non-trivial elements of p-power order, where p is prime, if and only if its order is divisible by p.*

Proof. Let A be a finite abelian group of order $p_1^{\gamma_1} \dots p_n^{\gamma_n}$ where $p_1, \dots, p_n$ are distinct primes. The theorem, together with the preceding remarks about p-groups, shows that A is the direct sum of groups of orders $p_1^{\gamma_1}, \dots, p_n^{\gamma_n}$. The corollary follows. □

COROLLARY 5.03. *If the order of a finite abelian group A is divisible by p^γ but not by $p^{\gamma+1}$ then A has a subgroup of order p^γ.* □

We remark that Theorem 5.01 shows that any finite abelian group contains a *unique* subgroup which is maximal in the sense that it is the set of all elements of p-power order, for each prime p; and that it is equally true that there is a unique subgroup containing precisely the elements of order prime to p. Such circumstances are generally absent, and are absent from the symmetric group S_3 in particular because the elements of 2-power order, which are also the elements with order prime to 3, do not form a subgroup.

Example 5.04. Consider the cyclic group C_m of order m where $m = p_1^{\alpha_1} \dots p_n^{\alpha_n}$. The theorem above states that C_m is the direct sum of cyclic subgroups of orders $p_1^{\alpha_1}, \dots, p_n^{\alpha_n}$ respectively. These may easily be found; if c generates C_m then clearly c^{m_i} generates the ith summand, where $m_i = m/p_i^{\alpha_i}$. Incidentally, this reasoning shows that the direct sum of cyclic groups of orders $p_1^{\alpha_1}, \dots, p_n^{\alpha_n}$ is isomorphic to the cyclic group of order $p_1^{\alpha_1} \dots p_n^{\alpha_n}$, provided of course that $p_1, \dots, p_n$ are distinct primes.

Theorem 5.01 shows that we need now consider only finite abelian p-groups in solving the problem in which we are primarily interested, namely to 'find' all finite abelian groups.

THEOREM 5.05. *A finite abelian p-group is isomorphic to the direct sum of cyclic p-groups.*

Proof. Let A be a finite abelian p-group. We have seen that A has order p^n for some non-negative integer n; and we use induction on n. When $n = 0$ there is, of course, nothing to prove, as $A = 0$, but this does start off the induction.

When $n > 0$ let b be any element of maximal order in A, and let $B = \text{gp}\{b\}$; we shall prove that B is a direct summand of A. If $B = A$ then A is itself cyclic and no further proof is necessary. We may therefore take it that both B and A/B have smaller order than A. By the inductive hypothesis, therefore, A/B is the direct sum of cyclic p-subgroups. Let generators of these subgroups be $a_1+B,..., a_r+B$, and let their orders be $p^{\alpha_1},..., p^{\alpha_r}$ where each α_i is positive. Thus $p^{\alpha_i}a_i \in B$ but $p^{\alpha_i-1}a_i \notin B$, for each i. It follows that $p^{\alpha_i}a_i = \beta_i b$ for certain integers β_i.

Showing that p^{α_i} divides β_i is the key step in the present proof. In order to derive a contradiction, therefore, suppose that we had $\beta_i = \beta_i' p^{\gamma_i}$ where β_i' is not divisible by p and $0 \leqslant \gamma_i < \alpha_i$. Thus
$$p^{\alpha_i}a_i = \beta_i' p^{\gamma_i} b$$
for $1 \leqslant i \leqslant r$. This equation asserts that the order of a_i exceeds that of b, because $\gamma_i < \alpha_i$, while of course $p^{\alpha_i}a_i$ and $p^{\gamma_i}b$ have the same order; observe that all elements of A have p-power order and that $p^{\alpha_i-1}a_i \neq 0$. But the order of b was to be greater than or equal to that of any element in A. Our assumption on β_i has led to a contradiction, so we must conclude that p^{α_i} divides β_i for each i.

Thus for each i there is an element b_i in B for which $p^{\alpha_i}a_i = p^{\alpha_i}b_i$. We put $c_i = a_i - b_i$, so that $p^{\alpha_i}c_i = 0$. We define C to be $\text{gp}\{c_1,..., c_n\}$ and we intend to prove that $A = B \oplus C$.

If x is an arbitrary element of A then $x+B \in A/B$ and $x+B$ has an expression $m_1(a_1+B)+...+m_r(a_r+B)$ for suitable integers $m_1,..., m_r$. Thus
$$x = m_1 a_1+...+m_r a_r+y$$
for some $y \in B$. Since $a_i = c_i+b_i$ we have
$$x = m_1 c_1+...+m_r c_r+z$$

where $z = m_1 b_1 + ... + m_r b_r + y \in B$, and so $x \in C+B$. Therefore $A = B+C$.

Next we suppose that $x \in B \cap C$, in the hope of showing that $x = 0$. Since $x \in C$ we have

$$x = m'_1 c_1 + ... + m'_r c_r$$

for suitable integers $m'_1, ..., m'_r$, so that

$$x+B = m'_1(c_1+B) + ... + m'_r(c_r+B).$$

Since $x \in B$ we have $x+B = B$. Recalling that $c_i = a_i - b_i$, we find that

$$B = m'_1(a_1+B) + ... + m'_r(a_r+B).$$

This implies that $m'_i(a_i+B) = B$ for each i, because A/B is the direct sum of cyclic subgroups, each generated by an a_i+B. Therefore $m'_i a_i \in B$, and so p^{α_i} divides m'_i. It follows (from $p^{\alpha_i} c_i = 0$) that $m'_i c_i = 0$, and the expression for x gives $x = 0$ at once.

We have now shown that $A = B \oplus C$, where B is cyclic, in view of Theorem 4.39. The inductive hypothesis shows that C is the direct sum of cyclic subgroups, and the theorem follows. □

It should be remarked that the proof of the preceding theorem is essentially constructive—a definite method is given for obtaining a direct decomposition in any specific case.

Example 5.06. Consider the multiplicative matrix group $\mathscr{G}$ generated by A and B where

$$A = \begin{pmatrix} i & 0 \\ 0 & i \end{pmatrix}, \qquad B = \begin{pmatrix} i & 0 \\ 0 & -i \end{pmatrix}.$$

This group is abelian because, as may be verified, $AB = BA$. (We shall use multiplicative notation in discussing it as additive notation would probably be even more confusing.) It will be found that $A^4 = I$ and $A^2 = B^2 = -I$ where I denotes the unit matrix, and that $\mathscr{G}$ has order 8. If $\text{gp}\{A\}$ is taken as one direct factor, as it may be, A being of order 4, then it can be verified that $\mathscr{G} \cong \text{gp}\{A\} \times \text{gp}\{AB\}$. This is not the only decomposition —we also have, for instance, $\mathscr{G} \cong \text{gp}\{B\} \times \text{gp}\{AB^{-1}\}$. (Note that $\mathscr{G}$ is *not* isomorphic to $\text{gp}\{A\} \times \text{gp}\{B\}$, which has order 16.) Though the decomposition is not unique, the orders of the direct summands are uniquely determined, as may (and later will quite generally) be proved.

We can now obtain the general result on finite abelian groups.

THEOREM 5.07. *A finite abelian group is isomorphic to the direct sum of a set of cyclic p-groups, for various primes p.*

Proof. By Theorems 5.01 and 5.05, the group is isomorphic to the direct sum of p-subgroups, for various primes p, and these p-subgroups are themselves isomorphic to direct sums of cyclic p-subgroups. □

The torsionfree abelian groups are our next consideration. The infinite cyclic group is torsionfree, and so indeed are all direct sums of groups each isomorphic to it.

LEMMA 5.08. *Let A be an abelian group generated by the elements $a_1,\ldots, a_n$, and suppose that there is a relation in A of the form*

$$m_1 a_1 + \ldots + m_n a_n = 0$$

where not all the integers $m_1,\ldots, m_n$ are 0. Then A has a set of generators $a_1',\ldots, a_n'$ such that $m'a_i' = 0$ for some i and for some non-zero integer m'.

Proof. We use induction on the integer $|m_1|+\ldots+|m_n|$, which will be denoted by m. By hypothesis $m > 0$; and when $m = 1$ just one $|m_i|$ is 1 and the rest are 0, and we take $a_j = a_j$ for $1 \leqslant j \leqslant n$.

Now suppose that $m > 1$. If $n-1$ of the numbers $m_1,\ldots, m_n$ are 0, then there is really nothing to prove. If not, then there must be two of them, say m_r and m_s with $r \neq s$, for which either $|m_r+m_s|$ or $|m_r-m_s|$ is less than $|m_r|$. We lose no generality in taking $r = 1$, $s = 2$, for this situation occurs when the generators $a_1,\ldots, a_n$ are suitably relabelled; and in taking $|m_1-m_2| < |m_1|$, for the proof with the other sign is only trivially different from what follows.

Since A is generated by $a_1,..., a_n$, it is also generated by $a_1, a_2+a_1, a_3,..., a_n$ (only the second generator has been changed). And the given relation in the given generating set is equivalent to

$$(m_1-m_2)a_1+m_2(a_2+a_1)+m_3a_3+...+m_na_n = 0.$$

We have $\quad |m_1-m_2|+|m_2|+|m_3|+...+|m_n| < m$

because $|m_1-m_2| < |m_1|$. The inductive hypothesis now applies: A has a set $\{a'_1,..., a'_n\}$ of generators and a relation $m'a'_i = 0$ (for some i) in them, with $m' \neq 0$. □

THEOREM 5.09. *A torsionfree finitely generated abelian group that is non-trivial is isomorphic to the direct sum of a set of infinite cyclic groups.*

Proof. Let the finitely generated abelian group A be torsionfree. It is tautologous that A has a finite generating set; but, of course, it has many such generating sets, and we wish to choose a special set $\{a_1,..., a_n\}$ such that $A \cong A_1 \oplus ... \oplus A_n$ where each A_i is generated by a_i.

Of all possible finite generating sets, we select any one with the minimal number of members, say n. In this case there is no relation of the form $m_1a_1+...+m_na_n = 0$ in A, unless all the m_i are taken as 0. For, otherwise, Lemma 5.08 shows that we can find a set $\{a'_1,..., a'_n\}$ of generators for A with the property that $m'a'_i = 0$ for some i and for m' non-zero. Since A is torsionfree we must have $a'_i = 0$, and then $\{a'_1,..., a'_{i-1}, a'_{i+1},..., a'_n\}$ is a generating set of $n-1$ elements. This contradiction to the choice of n shows that there is no relation in A of the form mentioned.

Each pair of elements taken from A_i, A_j respectively commute ($A_i = \text{gp}\{a_i\}$ for $1 \leqslant i \leqslant n$); $A = A_1+...+A_n$ as $\{a_1,..., a_n\}$ is a generating set; $A_i \cap (A_1+...+A_{i-1}+A_{i+1}+...+A_n) = 0$ because of the non-existence of relations having the form described above. The direct sum criterion of Theorem 4.39 therefore shows that $A = A_1 \oplus ... \oplus A_n$. Since A is torsionfree, each A_i is, of course, infinite cyclic. □

We now move on to the structure of those finitely generated abelian groups that are neither periodic nor torsionfree, the so-called mixed groups. Such objects exist—we have for example the direct sum of any (non-trivial) finite cyclic group and the infinite cyclic group. The following result applies to all abelian groups.

LEMMA 5.10. *The elements of finite order in an arbitrary abelian group A form a periodic subgroup P, and A/P is torsionfree.*

Proof. Note that A/P may be trivial—this will be the case if and only if A is periodic.

The fact and proof of the existence of a maximal periodic subgroup are very similar to what was seen earlier for maximal p-subgroups in a finite abelian group. The zero element 0 has finite order, so the set P of elements of finite order is non-empty. Let x, y be in P, so that $rx = 0$, $sy = 0$ for some positive integers r, s; then

$$(rs)(x-y) = s(rx)-r(sy) = s0-r0 = 0.$$

Hence $x-y \in P$. The subgroup criterion of Theorem 3.06 shows that P is a subgroup of A.

Suppose that A/P has an element $a+P$ of finite order m. Then $m(a+P) = P$, and $ma \in P$. Thus $ma = b$ where b has finite order. It follows that a itself has finite order, and so $a \in P$. Thus $a+P = P$. The only element in A/P with finite order, therefore, is P. Hence A/P is torsionfree. □

Example 5.11. The elements of infinite order will not in general form a subgroup. Let B be generated by an element b of finite order (greater than 1), and let C be generated by an element c of infinite order. The elements (b, c) and $(0, -c)$ of $B \oplus C$ have infinite order, but their sum $(b, 0)$ has neither infinite nor trivial order.

Another illusion is the hope that the elements of finite order in a non-abelian group form a subgroup. A counter-example will be found in Chapter 3, Problem 5.

THEOREM 5.12. *Let A be a finitely generated abelian group with periodic subgroup P. Then A is the direct sum of P and a suitable torsionfree subgroup T.*

Proof. By Lemma 5.10 A/P is torsionfree, and we may suppose that it is not trivial. Theorem 5.09 tells us that A/P, which is finitely generated, is isomorphic to the direct sum of cyclic groups. Let these be generated by $a_1+P,..., a_n+P$, respectively. Define T to be $\text{gp}\{a_1,..., a_n\}$. Then there can be no (non-trivial) relation of the form

$$m_1 a_1+...+m_n a_n = b$$

where $b \in P$. For, otherwise, we would have

$$m_1(a_1+P)+...+m_n(a_n+P) = P$$

as $b+P = P$; and the fact that A/P is the direct sum of infinite cycles generated by the a_i+P gives $m_i = 0$ for each i, and then $b = 0$ follows.

The subgroup T therefore has the property that $T \cap P = 0$. This implies that T is torsionfree. Another consequence is that $A \cong T \oplus P$, the least trivial fact in the application of the direct sum criterion of Theorem 4.39 now being the assertion that $A = T+P$, which follows from the definition of T. This completes the proof of the theorem. □

COROLLARY 5.13. *The periodic subgroup of a finitely generated abelian group is finitely generated.*

Proof. In the notation of the theorem $P \cong A/T$. As a factor group of a finitely generated group, P is also finitely generated. □

Note that whereas P is uniquely determined in the above theorem many choices for T are possible. Consider $B \oplus C$ of Example 5.11, in which P is precisely B. One choice for T is C; another is $\text{gp}\{(b, -c)\}$; in fact we may take $T = \text{gp}\{(nb, \pm c)\}$ where n is any integer.

The torsionfree subgroup T of the theorem is, however, determined *up to isomorphism*, for the reason that it is isomorphic to A/P and A itself determines the isomorphism class of A/P.

In general it is not true that the periodic subgroup of an abelian group is a direct summand. (A suitable counter-example will be presented after the necessary concepts have been developed; see Example 11.14.)

Our results thus far are now summarized.

THEOREM 5.14. *A finitely generated abelian group is isomorphic to the direct sum of a certain set of (finite, or infinite, or both finite and infinite) cyclic groups.*

Proof. If the abelian group A is finitely generated then $A \cong P \oplus T$ where P, T denote periodic, torsionfree subgroups respectively. Both P and T are finitely generated (recall Corollary 5.13). By Theorems 5.07 and 5.09 both P and T are isomorphic to direct sums of cyclic groups. The structure theorem for finitely generated abelian groups follows. □

Not every abelian group is isomorphic to the direct sum of cyclic groups. According to our current definition of direct sum such a group must be finitely generated, and even when a generalized definition of direct sum is used (as we shall see later) there are still abelian groups not isomorphic to the direct sum of cyclic groups; the additive group of rationals is such a group.

Though we now know the structure of all finitely generated abelian groups, we still do not know which of them are non-isomorphic. For all we know at the moment we might have $Z \oplus Z \oplus Z \cong Z \oplus Z$, where Z is the infinite cyclic group; or $Z_{p^3} \oplus Z_{p^3} \oplus Z_p \cong Z_{p^3} \oplus Z_{p^2} \oplus Z_{p^2}$, where Z_n is the cyclic group of order n. Our next aim, therefore, is to associate with each finitely generated abelian group a set of integers which determines completely its isomorphism class.

DEFINITION. An *invariant* of groups is a function (defined on groups) which takes the same value on isomorphic groups.

We proceed to prove formally that such obvious integers as the number of cyclic direct summands and their orders suffice to determine a finitely generated abelian group, up to isomorphism.

DEFINITION. The *rank* of a finitely generated abelian group is the number of infinite direct summands in any representation of the group as the direct sum of cyclic groups.

This definition has to be justified.

THEOREM 5.15. *The rank of a finitely generated abelian group is an invariant.*

Proof. Let P be the periodic subgroup of the finitely generated abelian group A; we have seen that A determines P uniquely. We know that $A \cong P \oplus T$ where T is a torsionfree subgroup of A, and T is determined up to isomorphism because $T \cong A/P$. It will therefore be enough to show that rank is an invariant of a torsionfree finitely generated abelian group.

Let S be the subset $\{2x : x \in T\}$ of T; that is, the set of squares (or rather 'doubles', in the additive notation) of elements in T. Since S is clearly a subgroup of T we may consider T/S. Suppose now that T has a representation as the direct sum of m infinite cyclic groups, generated by $t_1,\ldots, t_m$ respectively. We may, and shall, identify elements of T with elements $(\alpha_1 t_1,\ldots, \alpha_m t_m)$ of the direct sum, the α_i being integers, and in that case the elements of T/S have the form $(\alpha_1 t_1,\ldots, \alpha_m t_m)+S$ where each α_i is 0 or 1. The 2^m elements obtained in this way are distinct, which means that T/S has order 2^m. Of course, a representation of T as the direct sum of n infinite cyclic groups would imply in a similar way that T/S has order 2^n. Therefore $2^m = 2^n$ and $m = n$. This proves the invariance of the rank m of T. □

DEFINITION. The *type* of a finite abelian p-group isomorphic to the direct sum of a set of cyclic groups of orders $p^{n_1},\ldots, p^{n_r}$, where $n_1,\ldots, n_r$ are positive integers, is $\{n_1,\ldots, n_r\}$.

Justification is again needed.

THEOREM 5.16. *The type of a finite abelian p-group is an invariant.*

Proof. We suppose that the finite abelian p-group A has expressions both as the direct sum of cyclic groups generated by elements a_i of order p^{m_i} for $1 \leqslant i \leqslant r$, and as the direct sum of cyclic groups generated by elements b_i of order p^{n_i} for $1 \leqslant i \leqslant s$. We may assume that

$$m_1 \geqslant m_2 \geqslant \ldots \geqslant m_r,$$
$$n_1 \geqslant n_2 \geqslant \ldots \geqslant n_s;$$

and then we aim to prove that $r = s$ and that $m_i = n_i$ for $1 \leqslant i \leqslant r$.

We use induction on the order of A. The starting-point is the case when A has order p, and then there is no doubt that the type must be $\{1\}$.

In general, we take A of order greater than p. Which elements of A have order p or less? In the first direct sum representation they have the form $(\alpha_1 p^{m_1-1}a_1,\ldots, \alpha_r p^{m_r-1}a_r)$, each α_i taking the values $0,\ldots,p-1$. The elements just named are all distinct, and there are p^r of them. The second direct sum isomorphic to A implies that they number p^s. Since the number is certainly an invariant of A we have $p^r = p^s$, and $r = s$.

Next we consider $B = \{px : x \in A\}$. It is easy to see that B, which, of course, consists of certain multiples of elements of A, is a subgroup. In the first direct sum representation of A it may be identified with the elements of the form $(p\beta_1 a_1,\ldots, p\beta_r a_r)$ where $0 \leqslant \beta_i < p^{m_i-1}$ for $1 \leqslant i \leqslant r$. The elements just named are all distinct, because the subgroup they form is the direct product of the cyclic subgroups generated by $(0,\ldots, 0, pa_i, 0,\ldots, 0)$ for $1 \leqslant i \leqslant r$; the orders of these cyclic subgroups are $p^{m_1-1},\ldots, p^{m_r-1}$ respectively. In order to investigate the type of B we must remember that some m_i may be 1. Suppose that $m_i > 1$ for $1 \leqslant i \leqslant u$ and $m_i = 1$ for $u < i \leqslant r$; then the type of B is $\{m_1-1,\ldots, m_u-1\}$.

The second direct sum representation of A gives another for B. Suppose that $n_i > 1$ for $1 \leqslant i \leqslant v$ and $n_i = 1$ for $v < i \leqslant r$; then a similar argument shows that the type of B is $\{n_1-1,\ldots, n_v-1\}$.

We may use the induction hypothesis here, B having smaller order than A. We conclude firstly that $u = v$; and secondly that

$$m_1 - 1 = n_1 - 1, \quad \ldots, \quad m_u - 1 = n_u - 1.$$

Thus $m_i = n_i$ for $1 \leqslant i \leqslant u$, while $m_i = 1$ and $n_i = 1$ for $u < i \leqslant r$. The two proposed types of A are therefore the same, and the inductive proof that the type of A is an invariant of A is complete. □

Note that a finite abelian p-group of type $\{n_1, \ldots, n_r\}$ has order p^n where $n = n_1 + \ldots + n_r$. Groups in which each n_i is 1 are called *elementary abelian p-groups*; they are just isomorphic copies of direct sums of groups of order p, or just the abelian groups in which every element has order p or 1. We proved (in Chapter 4) that any group with each element of order 2 or 1 must be abelian, and now we see that (if it is finite) it is isomorphic to the direct sum of groups of order 2.

We can now give a complete description of the groups under study.

THEOREM 5.17. *A finitely generated abelian group determines, and is determined up to isomorphism by, the following invariants: the rank, the (finite) set of primes p occurring as orders of elements, and the type (for each p) of the subgroup formed by the elements of p-power order.*

Proof. The last two theorems, together with the facts that the periodic subgroup and the p-subgroups are unique finite subgroups, show that the group determines the invariants mentioned. Now suppose that a set of invariants is given—rank q, set $\{p_1, \ldots, p_r\}$ of primes, and a type for each prime. If type $\{n_1, \ldots, n_k\}$ corresponds to the typical prime p in the given set, we construct the direct sum of cyclic groups of orders $p^{n_1}, \ldots, p^{n_k}$ respectively; we do this for each p, take the direct sum of all the resulting groups, and take the direct sum of this with q infinite cyclic groups. The outcome is a finitely generated abelian group with the given set of invariants. □

We describe briefly another set of invariants of the finitely generated abelian group A. The periodic subgroup P of A is the direct sum of groups of order $p_i^{m_i}$ where $1 \leqslant i \leqslant r$ and $m_i > 0$, while each p_i-subgroup is the direct sum of cyclic groups of orders $p_i^{n_{i1}}, p_i^{n_{i2}}, \ldots, p_i^{n_{ik}}$, say; here we allow some of the n_{ij} to be 0, but we still suppose that each $n_{ij} > 0$ for at least *one* value of i. We may without significant loss of generality assume that

$$p_1 > p_2 > \ldots > p_r,$$

and that

$$n_{i1} \geqslant n_{i2} \geqslant \ldots \geqslant n_{ik}$$

for each i. Let us write the orders of our direct summands of P as a rectangular array the better to visualize the position:

$$\begin{array}{cccccc} p_1^{n_{11}} & p_1^{n_{12}} & \cdot & \cdot & \cdot & p_1^{n_{1k}} \\ p_2^{n_{21}} & p_2^{n_{22}} & \cdot & \cdot & \cdot & p_2^{n_{2k}} \\ \cdot & \cdot & \cdot & \cdot & \cdot & \cdot \\ p_r^{n_{r1}} & p_r^{n_{r2}} & \cdot & \cdot & \cdot & p_r^{n_{rk}}. \end{array}$$

Now the cyclic direct summands of orders $p_1^{n_{11}}, \ldots, p_r^{n_{r1}}$ generate a cyclic subgroup of order $p_1^{n_{11}} \ldots p_r^{n_{r1}}$; in general, the cyclic direct summands of orders $p_1^{n_{1j}}, \ldots, p_r^{n_{rj}}$ generate a cyclic subgroup of order $p_1^{n_{1j}} \ldots p_r^{n_{rj}}$ for $1 \leqslant j \leqslant k$, because the primes $p_1, \ldots, p_r$ are distinct. Put $d_j = p_1^{n_{1j}} \ldots p_r^{n_{rj}}$. Our assumptions imply that d_j is divisible by d_{j+1} for $1 \leqslant j < k$. It is easy to see that P is isomorphic to the direct sum of cyclic groups of orders $d_1, \ldots, d_k$; and that P may be (and sometimes is) specified by these numbers, which are invariants of A.

Although we now have a rather comprehensive description of the finitely generated abelian groups, it does not suffice to answer all possible questions about the structure of such groups. The following theorem, the proof offered for which is independent of the theory so far developed, answers one question of this sort.

THEOREM 5.18. *If an abelian group is generated by n elements then every subgroup of it is generated by n elements.*

Proof. We use induction on n. When $n = 1$ the group in question is cyclic, and Theorem 3.21 informs us that every subgroup is also cyclic.

We suppose therefore that the abelian group A is generated by n elements, with $n > 1$. Note that we are not assuming that n is in any way minimal, as we might if (for instance) we wished to apply our structure theory. Let $a_1,\ldots, a_n$ be some set of generators; then each element of the subgroup B has some expression as $\beta_1 a_1+\ldots+\beta_n a_n$ where $\beta_1,\ldots, \beta_n$ are suitable integers. If it should happen that *every* element of B has such an expression with $\beta_1 = 0$, then $B \leqslant \text{gp}\{a_2,\ldots, a_n\}$; and in this case the inductive hypothesis ensures that B is generated by some set of $n-1$ of its elements. It certainly follows that B is generated by n elements.

In the general case there will be an element, b say, in B with an expression of the above form in which $\beta_1 \neq 0$; replacing b by $-b$ if need be, we may take it that $\beta_1 > 0$. We make a definite choice of b; we take b as any element of B for which there is a representation as $\beta_1 a_1+\ldots+\beta_n a_n$ in which β_1 takes its minimal positive value. Take next a general element x of B. It will have some expression of the form

$$x = \gamma_1 a_1+\ldots+\gamma_n a_n.$$

The usual application of the division algorithm gives

$$\gamma_1 = q\beta_1+r, \quad 0 \leqslant r < \beta_1.$$

The element $x-qb$ lies in B, and we have

$$\begin{aligned} x-qb &= (\gamma_1 a_1+\ldots+\gamma_n a_n)-q(\beta_1 a_1+\ldots+\beta_n a_n) \\ &= ra_1+(\gamma_2-q\beta_2)a_2+\ldots+(\gamma_n-q\beta_n)a_n. \end{aligned}$$

But β_1 had a certain minimal property, which tells us that since $0 \leqslant r < \beta_1$ we must have $r = 0$; that is, $x-qb \in \text{gp}\{a_2,\ldots, a_n\}$. The significance of this is that $B = \text{gp}\{b\}+(\text{gp}\{a_2,\ldots, a_n\} \cap B)$.

At this point the inductive hypothesis may be applied. Therefore the elements of B which lie in $\text{gp}\{a_2,\ldots, a_n\}$ form a subgroup with a generating set of $n-1$ elements. Since B is the sum of this subgroup and $\text{gp}\{b\}$, it follows that B has a set of n generators. □

Problems

1. Express as the direct product of cyclic subgroups that subgroup of the multiplicative group of non-zero rationals generated by $\{6, -6\}$.

2. Show that the multiplicative group generated by the matrices A and B is abelian if

$$A = \begin{pmatrix} 1 & 1 \\ -1 & 1 \end{pmatrix}, \qquad B = \begin{pmatrix} -1 & 1 \\ -1 & -1 \end{pmatrix}.$$

Express this group as the direct product of cyclic subgroups in two ways.

3. Show that the permutations

$$(1,2,3,4)(5,6,7,8) \quad \text{and} \quad (1,5,3,7)(2,6,4,8)(9,10,11)$$

generate an abelian subgroup A of the symmetric group S_{11}, and that A has order 24. Express A as the direct product of cyclic subgroups.

4. Express all the abelian groups of orders 99 and 100 as direct sums of cyclic groups.

5. How many (non-isomorphic) abelian groups of order 10^6 are there? How many of them are cyclic?

6. The infinite cyclic groups A, B are generated by the elements a, b respectively. Find (in terms of a and b) all pairs of elements which generate $A \oplus B$.

7. Find all finitely generated abelian groups A that have no characteristic subgroups other than A and 0.

8. Let A be the abelian group formed by taking the direct sum of cyclic groups of orders $d_1, \ldots, d_k$. Prove that, if d_{j+1} divides d_j for $1 \leqslant j < k$ and if $d_k > 1$, then no set of $k-1$ elements of A can generate A.

9. The abelian group A is defined to be $A_1 \oplus \ldots \oplus A_n$ where $A_i = \mathrm{gp}\{a_i\}$ for $1 \leqslant i \leqslant n$. The subgroup B is defined to be $\mathrm{gp}\{b\}$, where

$$b = a_1 + \beta_2 a_2 + \ldots + \beta_n a_n$$

for certain constants $\beta_2, \ldots, \beta_n$. Prove that

$$A = B \oplus A_2 \oplus \ldots \oplus A_n$$

whenever a_1 has infinite order; and that if a_1 has finite order m then

$$A = B \oplus A_2 \oplus \ldots \oplus A_n$$

if and only if $m\beta_2 a_2 = 0, \ldots, m\beta_n a_n = 0$.

10. The abelian group A is generated by elements a and b for which $2a = 3b$. Express A as the direct sum of (not necessarily non-trivial) cyclic groups.

11. Prove that, if B is a subgroup of the finitely generated abelian group A, then the rank of A equals the sum of the ranks of B and A/B.

12. Let A denote the direct sum of the cyclic groups generated by the elements a_1, a_2, a_3, of orders p^3, p^3, p^2 respectively (p is a prime), and let $B = \text{gp}\{(pa_1, p^2a_2, 0), (0, pa_2, pa_3)\}$. Find elements b_1, b_2, b_3 and constants $\beta_1, \beta_2, \beta_3$ such that

$$A = \text{gp}\{b_1\} \oplus \text{gp}\{b_2\} \oplus \text{gp}\{b_3\},$$
$$B = \text{gp}\{\beta_1 b_1\} \oplus \text{gp}\{\beta_2 b_2\} \oplus \text{gp}\{\beta_3 b_3\}.$$

13. In the notation and situation of the previous question, let $C = \text{gp}\{pa_1, (p^2+p)a_2, pa_3\}$. Can the b_i and the β_i be so chosen that, in addition,

$$C = \text{gp}\{\gamma_1 b_1\} \oplus \text{gp}\{\gamma_2 b_2\} \oplus \text{gp}\{\gamma_3 b_3\}$$

for some constants γ_1, γ_2, γ_3?

14. Determine which of the following statements about finitely generated abelian groups are true and which are false.

(i) Every factor group of a torsionfree group is torsionfree.

(ii) Every infinite, finitely generated abelian group has a set of generators, each of which has infinite order.

(iii) If A_1, A_2 are subgroups of the group A with $A_1 \cong A_2$ then $A/A_1 \cong A/A_2$.

(iv) If A_1, A_2 are subgroups of the group A with $A/A_1 \cong A/A_2$, then $A_1 \cong A_2$.

(v) Every finitely generated abelian group has a subgroup of prime order.

15. Let the abelian group A be generated by its elements $a_1, \ldots, a_n$, and let $m_1, \ldots, m_n$ be given integers with highest common factor 1. Show, by induction on $|m_1| + \ldots + |m_n|$ or otherwise, that there exists a set $b_1, \ldots, b_n$ of elements generating A, with

$$b_1 = m_1 a_1 + \ldots + m_n a_n.$$

**16. Let the abelian group A be generated by a set of k elements, and choose a set $\{a_1, \ldots, a_k\}$ of generators with the following properties:

(i) the order of a_i is no less than the order of a_{i+1} for $1 \leqslant i < k$ (infinite order is taken to exceed all finite orders);

(ii) there is no generating set $\{b_1, \ldots, b_k\}$ which, when similarly arranged, is such that for some $m < k$ the orders of the first m pairs $\{a_i, b_i\}$ are the same, while the order of b_{m+1} is less than that of a_{m+1}.

By using the result of the preceding problem or otherwise, prove that $A \cong \text{gp}\{a_1\} \oplus \ldots \oplus \text{gp}\{a_k\}$.

17. An abelian group A is said to be *divisible* if the equation $nx = a$ has a solution x for all positive integers n and for all $a \in A$. Prove that no finitely generated abelian group is divisible. Show that the additive group

Q of rationals, and the group Z_p^∞ (see Example 1.09), are each divisible. Hence, or otherwise, prove that neither of these groups is finitely generated.

18. Let $\mathrm{Hom}(A, C^*)$ denote the set of homomorphisms from the finite abelian group A into the multiplicative group C^* of non-zero complex numbers (see Problem 3 of Chapter 2).

(i) Show that $\mathrm{Hom}(A, C^*)$ is an abelian group. Express it as the direct sum of prime-power cyclic groups, given such a decomposition for A.

(ii) Prove that A is isomorphic to $\mathrm{Hom}(A, C^*)$ in a natural way.

(iii) Prove that, if a in A corresponds to χ_a under this isomorphism, then

$$b\chi_a = a\chi_b$$

for all a, b in A.

6

Normal structure

THE classical theory of finite groups (apart from the theory of group representations) rests on two pillars—normal structure and Sylow structure; the former is concerned with properties defined by normal subgroups and factor groups, the latter with the existence of certain subgroups of a group and arithmetic properties involving the orders of these subgroups. The present chapter and the next are therefore devoted to basic ideas of normal structure and Sylow structure. Though finite groups are the prime object of study we do not exclude infinite groups except when we find it absolutely necessary.

Our first results are elementary technical theorems about normal subgroups and factor groups, indeed they are little more than collections of lemmas, but they are none the less essential to our progress.

THEOREM 6.01. *Let U, V, S be subgroups of the group G such that U is a normal subgroup of V. Then $S \cap U$ is a normal subgroup of $S \cap V$.*

Proof. Clearly $S \cap U$ is a subgroup of S, and it is contained in $S \cap V$ since $U \leqslant V$; it follows that $S \cap U$ is a subgroup of $S \cap V$.

Take arbitrary elements u, v in $S \cap U$, $S \cap V$ respectively, and consider $v^{-1}uv$. Since $U \trianglelefteq V$ we have $v^{-1}uv \in U$; since $u, v \in S$ we have $v^{-1}uv \in S$. Hence $v^{-1}uv \in S \cap U$. It follows that

$$v^{-1}(S \cap U)v \subseteq S \cap U$$

for all $v \in S \cap V$. This suffices to show that $S \cap U$ is normal in $S \cap V$. □

COROLLARY 6.02. *If U is a normal subgroup, and S is any subgroup, of the group G, then $S \cap U$ is a normal subgroup of S.*

Proof. Take $V = G$ in the theorem. □

COROLLARY 6.03. *If U is a normal subgroup of G and if U is contained in the subgroup S of G then U is a normal subgroup of S.*

Proof. Apply Corollary 6.02. □

Having studied normal subgroups in relation to subgroups of a given group, we turn to the effect on subgroups of taking a certain kind of factor group of a given group.

THEOREM 6.04. *If N is a normal subgroup of the group G and if the subgroup A of G contains N, then A/N is a subgroup of G/N. If in addition A is normal in G then A/N is normal in G/N.*

Conversely, if N is a normal subgroup of the group G, then each subgroup of G/N has the form A/N for a suitable subgroup A of G. In addition each normal subgroup of G/N has the form A/N for a suitable normal subgroup A of G.

Proof. If $N \trianglelefteq G$ and $N \leqslant A \leqslant G$ then Corollary 6.03 shows that $N \trianglelefteq A$. Therefore A/N exists. An arbitrary element of A/N has the form Na where $a \in A$, and so $a \in G$; therefore $Na \in G/N$, and $A/N \subseteq G/N$. We prove that $A/N \leqslant G/N$ by the usual subgroup criterion, Theorem 3.06. First, A/N is non-empty as $N \in A/N$. Second, if Na, $Nb \in A/N$ then $(Na)(Nb)^{-1} = Nab^{-1} \in A/N$ since $ab^{-1} \in A$; of course, $a, b \in A$ and A is a subgroup. Therefore $A/N \leqslant G/N$.

Now let $A \trianglelefteq G$; thus $g^{-1}ag \in A$ for all a, g in A, G respectively. Arbitrary elements of A/N, G/N have the forms Na, Ng respectively.

$$(Ng)^{-1}(Na)(Ng) = Ng^{-1}ag \in A/N$$

because $g^{-1}ag \in A$. Therefore $(Ng)^{-1}(A/N)(Ng) \subseteq A/N$ for all Ng in G/N, and this suffices to establish that A/N is normal in G/N.

Next, the converse. Although this could easily be proved by arguments like those above, we prefer to rely on the homomorphism concept for the sake of simplicity and variety.

Since N is normal in G, the group G/N exists; call it H. There is the natural homomorphism ϕ from G to H, with kernel N—it simply maps x onto Nx. Let K be any subgroup of H. Define A to be the subset $\{a : a\phi \in K\}$ of G. We shall prove that A is the subgroup of G which we seek.

Clearly $1 \in A$. If $a, b \in A$ then

$$(ab^{-1})\phi = (a\phi)(b\phi)^{-1} \in K$$

since $a\phi, b\phi \in K$, and K is known to be a subgroup of H. By the usual subgroup criterion A is a subgroup of G. The restriction of ϕ to A is, of course, a homomorphism of A into H and indeed onto K; and the kernel of ϕ is, of course, N. By the first isomorphism theorem (4.07), $A/N \cong K$, as required.

Suppose, finally, that K is normal in H. We have to show that $A \trianglelefteq G$. Take arbitrary elements a, g in A, G respectively, and consider $g^{-1}ag$. Since A/N is normal in G/N we have

$$(Ng)^{-1}(Na)(Ng) \in A/N,$$

that is, $Ng^{-1}ag \in A/N$. Thus $g^{-1}ag \in A$,

$$g^{-1}Ag \subseteq A.$$

This suffices to prove A normal in G. □

COROLLARY 6.05. *Let N be a normal subgroup of the group G. Then there is a one–one correspondence between subgroups of G containing N and subgroups of G/N; and in it normal subgroups correspond to normal subgroups.*

Proof. Let $N \leqslant A \leqslant G$. The procedure of the theorem was to map A to A/N, and to associate with each subgroup K of G/N a subgroup A of G containing N and satisfying $K = A/N$. The proof makes it clear that if $A/N = B/N$ then $A = B$. It is trivial that $A = B$ implies $A/N = B/N$. The desired one–one correspondence therefore exists. The statement about normal subgroups should now be obvious. □

Our next task is again routine: we ask if there are operations for normal subgroups analogous to set-theoretic intersection and union.

THEOREM 6.06. *Let $\{N_\lambda : \lambda \in \Lambda\}$ be a set of normal subgroups of the group G, where Λ is some index set. Then $\bigcap_{\lambda \in \Lambda} N_\lambda$ is a normal subgroup of G.*

Proof. Theorem 3.07 implies that if $N = \bigcap_{\lambda \in \Lambda} N_\lambda$ in these circumstances then N is a subgroup of G. Take arbitrary elements g, n in G, N respectively. Since $n \in N_\lambda$ for all $\lambda \in \Lambda$, and since $N_\lambda \trianglelefteq G$, we have

$$g^{-1}ng \in N_\lambda.$$

Since this holds for *all* $\lambda \in \Lambda$ we deduce that $g^{-1}ng \in N$; that is, $g^{-1}Ng \subseteq N$. This suffices to prove that N is normal in G. □

An important aspect of this theorem is that it gives a meaning to the expression 'the normal subgroup generated by the complex C of the group G'. We may define this meaning to be 'the intersection of all normal subgroups of G containing C', the theorem guaranteeing that this intersection is itself normal in G. It may be regarded as the least normal subgroup in which C lies.

Example 6.07. We determine the normal subgroup N generated by the complex $\{(12)\}$ in the symmetric group S_n of degree $n \geqslant 2$. Let $a = (12)$, $b = (12\ldots n)$. An easy calculation gives $b^{-i}ab^i = (i+1, i+2)$ for $i = 0, 1, \ldots, n-2$; see Problem 10 of Chapter 3; therefore $(i+1, i+2) \in N$ for such i. But every element of S_n can be expressed as a product of such elements. Therefore $N \geqslant S_n$, and we find that $N = S_n$.

The reader may be able to find a description (which we omit) for the normal subgroup generated by a complex, along the lines of Theorem 3.10.

We saw in Chapter 3 that unions of subgroups are not usually subgroups, and unions of normal subgroups are no more interesting to study. As the subgroups *generated* by a set of normal subgroups and the complex product of the latter are closely related, we shall analyse this situation in detail and with care.

THEOREM 6.08. *Let A and B be subgroups of the group G. Then AB is a subgroup of G if and only if $AB = BA$.*

Proof. Suppose $AB = BA$ where, of course, AB is the complex product of A and B. Since A and B are subgroups AB is certainly non-empty; $1 \in AB$. Take two elements of AB, say $a_i b_i$ for $i = 1, 2$ (with $a_i \in A, b_i \in B$). We consider, as usual, their quotient $(a_1 b_1)(a_2 b_2)^{-1}$, which is $a_1 b_1 b_2^{-1} a_2^{-1}$. Since $BA = AB$ there are elements a, b in A, B respectively for which

$$(b_1 b_2^{-1}) a_2^{-1} = ab.$$

Thus $(a_1 b_1)(a_2 b_2)^{-1} = a_1 ab \in AB$. It follows from the subgroup criterion of Theorem 3.06 that AB is a subgroup of G.

Conversely, suppose that AB is a subgroup of G. An arbitrary element of BA has the form ba, with $a \in A$, $b \in B$. Then $(ba)^{-1} = a^{-1}b^{-1} \in AB$, because A, B are subgroups; hence $ba \in AB$ because AB is a subgroup. Therefore $BA \subseteq AB$.

We cannot claim here that $AB \subseteq BA$ 'similarly', because we do not know that BA is a subgroup. However, the proof is not difficult. Take an element x in AB. Then $x^{-1} \in AB$ (which is a subgroup), so x^{-1} has some expression as ab, with $a \in A$, $b \in B$. But then $x = b^{-1}a^{-1} \in BA$ since A, B are subgroups. We have thus proved that $AB \subseteq BA$.

It follows that $AB = BA$, as required. □

Only the former half of the above theorem is strictly relevant to our applications.

COROLLARY 6.09. *If A and B are subgroups of the group G such that $AB = BA$ then* $\text{gp}\{A, B\} = AB$.

Proof. By the theorem AB is a subgroup, and as $AB \supseteq A$, $AB \supseteq B$, we have $AB \supseteq \text{gp}\{A, B\}$. But, on the other hand, $\text{gp}\{A, B\}$ certainly contains A and B and so all elements ab (where $a \in A$, $b \in B$); thus $\text{gp}\{A, B\} \supseteq AB$. It follows that $\text{gp}\{A, B\} = AB$. □

COROLLARY 6.10. *If A is a subgroup of the group G and N is a normal subgroup of G then* $\text{gp}\{A, N\} = AN$.

Proof. The definition of a normal subgroup N implies that $xN = Nx$ for all x in G. It certainly follows that $AN = NA$ for all subgroups A of G. Corollary 6.09 now gives the desired result. □

THEOREM 6.11. *If N_1, N_2 are normal subgroups of the group G then $N_1 N_2$ is also a normal subgroup.*

Proof. The previous theorem and its corollaries make it abundantly clear that $N_1 N_2$ is a subgroup of G. Normality may be shown directly from the definition: if x is any element of G then $x^{-1}N_i x = N_i$ for $i = 1, 2$, and so

$$x^{-1}N_1 N_2 x = x^{-1}N_1 x . x^{-1}N_2 x = N_1 N_2,$$

as required. □

We turn now from the study of normal subgroups to study of factor groups. The need for the two theorems which follow may best be felt in terms of homomorphisms, for they answer these questions:

(i) If A is a subgroup of G and ϕ is some homomorphism of G with kernel N, what (in terms of G/N) is the image of A under ϕ regarded as a homomorphism of A?

(ii) If homomorphisms with given kernels map G onto G_1 and G_1 onto G_2 respectively, how are G, G_1, G_2 related?

Theorem 6.04, which showed that if $N \leqslant A$ then A/N is a subgroup of G/N, is related to the following answer to (i):

THEOREM 6.12 (Second isomorphism theorem). *If A is a subgroup and N is a normal subgroup of the group G then $A/(A \cap N) \cong NA/N$.*

Proof. Note that, by Corollaries 6.02, 6.10, and 6.03, $A \cap N \trianglelefteq A$, NA is a subgroup of G, and $N \trianglelefteq NA$.

The idea of the proof is to define a homomorphism from A onto NA/N, to determine its kernel K, and to assert that $A/K \cong NA/N$ by the first isomorphism theorem.

We define a mapping ϕ from A to NA/N by putting $a\phi = Na$. The general element of NA/N has the form Nna or Na (here $n \in N$); so ϕ is onto NA/N. Clearly ϕ is a homomorphism:

$$a\phi . b\phi = Na . Nb = Nab = (ab)\phi.$$

If $x \in K$ then $x\phi = N$, because N is the identity element of both G/N and NA/N. But $Nx = N$ means precisely that $x \in N$. Of course ϕ is a mapping of A, so $x \in A \cap N$. Thus $K \leqslant A \cap N$. Conversely, if $y \in A \cap N$ then $y\phi$ is defined and $y \in N$ shows that $y\phi = 1$, so $y \in K$. It follows that $K = A \cap N$.

By the first isomorphism theorem, then,

$$A/(A \cap N) \cong NA/N. \square$$

Example 6.13. Take G to be the symmetric group S_4, in which N is to be gp$\{(12)(34), (13)(24)\}$ and A is to be gp$\{(1234)\}$. A Venn diagram to describe the situation may be permitted:

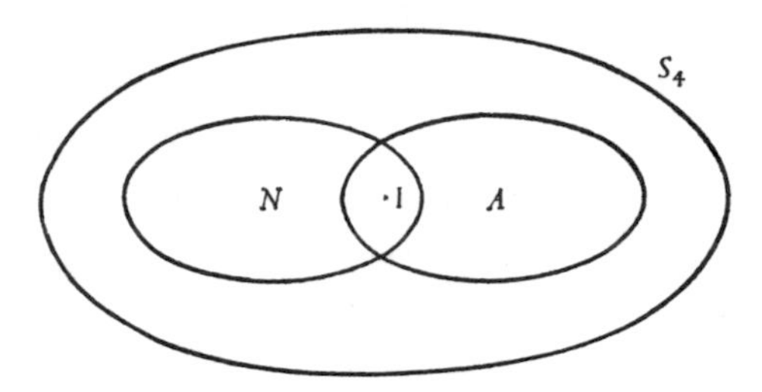

It should be verified that $N \trianglelefteq S_4$ and that $A \cap N = \{1, (13)(24)\}$. Thus $A \cap N$ has index 2 in A, and $A/(A \cap N)$ consists of these two cosets:

$$A \cap N = \{1, (13)(24)\},$$
$$(A \cap N)(1234) = \{(1234), (1432)\}.$$

On the other hand NA consists of eight elements:

$$\{1, (12)(34), (13)(24), (14)(23), (1234), (13), (1432), (24)\}.$$

It follows that NA/N has just two elements:

$$N = \{1, (12)(34), (13)(24), (14)(23)\},$$
$$N(1234) = \{(1234), (13), (1432), (24)\}.$$

Just as the theorem predicted, we find that $A/(A \cap N) \cong NA/N$. The homomorphism ϕ of the theorem would map (1234) in A onto $N(1234)$, so in the final isomorphism $(A \cap N)(1234)$ would correspond to $N(1234)$.

A lemma will precede our answer to question (ii) above. It is unnecessarily general for the task in hand, but significant in numerous other contexts (such as the proof of Theorem 4.07).

LEMMA 6.14. *If ϕ is a homomorphism from the group G_1 into the group G_2 which maps N_1 into N_2 (N_1, N_2 being normal subgroups of G_1, G_2 respectively) then there is a homomorphism ψ from G_1/N_1 into G_2/N_2, defined by*

$$(N_1 x)\psi = N_2(x\phi)$$

for all x in G_1.

Proof. It is far from trivial that ψ is a mapping—it could conceivably happen that $N_2(x\phi)$ depends essentially on the representative x chosen from the coset $N_1 x$. We therefore have to show that if $N_1 x = N_1 y$ then $N_2(x\phi) = N_2(y\phi)$. From $N_1 x = N_1 y$ we have $y = nx$ for some $n \in N_1$; so

$$y\phi = (nx)\phi = (n\phi)(x\phi) \in N_2(x\phi)$$

because ϕ is a homomorphism. It follows that $N_2(y\phi) = N_2(x\phi)$, by Theorem 3.12, as required.

To prove that ψ is a homomorphism is now straightforward. If $g, h \in G_1$, then

$$\begin{aligned}(N_1 g)\psi . (N_1 h)\psi = N_2(g\phi) . N_2(h\phi) &= N_2(g\phi)(h\phi) \\ &= N_2\{(gh)\phi\} \\ &= (N_1 gh)\psi\end{aligned}$$

since ϕ is a homomorphism, and the lemma is proved. □

THEOREM 6.15 (Third isomorphism theorem). *If M and N are normal subgroups of the group G such that M contains N then $(G/N)/(M/N) \cong G/M$.*

Proof. Note that we have evidence (in Corollary 6.03 and Theorem 6.04) to assert that M/N exists and that it is a normal subgroup of G/N.

The idea of the proof is to define a homomorphism ψ from G/N to G/M, to find its kernel, and again to appeal to the first isomorphism theorem.

We try to define ψ by

$$(Nx)\psi = Mx$$

for all $x \in G$, and Lemma 6.14 (with $G_1 = G, N_1 = N, G_2 = G/M$,

$N_2 = 1$, and ϕ the natural homomorphism from G to G/M) shows that this is indeed a definition of a homomorphism.

The kernel is a normal subgroup of G/N and (by Theorem 6.04) has the form K/N where K is some normal subgroup of G. Let $Nx \in K/N$, then $(Nx)\psi = M$. But $Mx = M$ is equivalent to $x \in M$, and so $K/N \leqslant M/N$. Since it is clear that if $x \in M$ then $(Nx)\psi = M$ and so $Nx \in K/N$, it follows that $K/N = M/N$.

The first isomorphism theorem therefore shows that

$$(G/N)/(M/N) \cong G/M. \square$$

Example 6.16. Let G be Z, the additive group of integers, and let r, s be two positive integers. There is a homomorphism ϕ_{rs} of Z onto the group Z_{rs} of the residue classes modulo rs defined by $a\phi_{rs} = [a]$. Its kernel is N, consisting of all integral multiples of rs. Thus $Z/N \cong Z_{rs}$. Now Z_{rs} in turn has a homomorphism ϕ_r onto the group Z_r of the residue classes modulo r, defined by $[a]\phi_r = [a]$; note the two meanings of $[a]$. The kernel of ϕ_r is $\{[a] : a \text{ is a multiple of } r\}$. This is a subgroup of Z_{rs}, and corresponds to a subgroup M/N of Z/N. The theorem above confirms that M consists of all multiples of r.

Example 6.17. We consider a homomorphism ϕ from the group G onto an abelian group, with particular reference to Theorem 4.21. Thus the kernel M contains $\delta(G)$. By Theorem 6.15 we have

$$G/M \cong \{G/\delta(G)\}/\{M/\delta(G)\}.$$

In terms of homomorphisms, this states that $\phi = \phi_1\phi_2$ where ϕ_1 is the natural homomorphism of G onto $G/\delta(G)$ and ϕ_2 is the natural homomorphism of $G/\delta(G)$ onto $\{G/\delta(G)\}/\{M/\delta(G)\}$. Thus every homomorphism of G onto an abelian group may be expressed as the product of ϕ_1 and a suitable homomorphism of $G/\delta(G)$.

We now approach the crucial result on normal structure of finite groups, Theorem 6.23 below due to Jordan, Hölder, and others. The preparations for this include a result which is traditionally known as Zassenhaus's lemma but which is in fact a fourth isomorphism theorem:

THEOREM 6.18. *Let A, B be subgroups of the group G and let X, Y be normal subgroups of A, B respectively. Then*

$$X(A \cap B)/X(A \cap Y) \cong Y(A \cap B)/Y(X \cap B).$$

Proof. The statement of the theorem is meant to include implicitly the fact that the required factor groups exist.

We shall prove that $X(A \cap B)/X(A \cap Y)$ is isomorphic to $(A \cap B)/(A \cap Y)(X \cap B)$. This will suffice, for the latter factor group, which is unaltered when X, A are interchanged with Y, B respectively, is then clearly isomorphic to $Y(A \cap B)/Y(X \cap B)$ also; the required result follows.

Because $A \cap B$ is a subgroup, and X is a normal subgroup, of A, we see (from Corollary 6.10) that $X(A \cap B)$ is a subgroup of A. But X is normal in $X(A \cap B)$, and we can meaningfully consider $X(A \cap B)/X$. The second isomorphism theorem gives

$$X(A \cap B)/X \cong (A \cap B)/(A \cap B \cap X) = (A \cap B)/(X \cap B).$$

Now $(A \cap B)/(A \cap Y)(X \cap B)$ is a factor group of $(A \cap B)/(X \cap B)$, for quite generally if we have N_1, $N_2 \trianglelefteq G$ then $N_1 N_2 \trianglelefteq G$, $N_2 \trianglelefteq N_1 N_2$, and

$$G/N_1 N_2 \cong (G/N_2)/(N_1 N_2/N_2)$$

by the third isomorphism theorem; of course both $A \cap Y$ and $X \cap B$ are normal in $A \cap B$, by Theorem 6.01. There is, therefore, a homomorphism from $X(A \cap B)/X$ onto $(A \cap B)/(A \cap Y)(X \cap B)$.

Our tactics now are to determine the kernel, K/X say, of this homomorphism and then to appeal to the first isomorphism theorem. If Xk lies in the kernel, with $k \in A \cap B$, then $k \in (A \cap Y)(X \cap B)$, and it follows that $K \leqslant X(A \cap Y)$ as we are assuming that $K \geqslant X$. Conversely, take $k \in X(A \cap Y)$, then $k \in X(A \cap B)$, and the image of Xk is clearly trivial. Hence $K/X = X(A \cap Y)/X$.

The first isomorphism theorem shows that

$$\{X(A \cap B)/X\}/\{X(A \cap Y)/X\} \cong (A \cap B)/(A \cap Y)(X \cap B).$$

After simplifying the left-hand side by an application of the third isomorphism theorem we obtain

$$X(A \cap B)/X(A \cap Y) \cong (A \cap B)/(A \cap Y)(X \cap B).$$

This, as was pointed out above, leads to the assertion made in the theorem. □

Preparatory definitions are also necessary. All series of subgroups to be considered, it should be noted, will be *finite* series.

DEFINITION. A *normal series* of a group G is a finite series of subgroups

$$G = G_0 \trianglerighteq G_1 \trianglerighteq \ldots \trianglerighteq G_{r-1} \trianglerighteq G_r = 1$$

such that G_i is a normal subgroup of G_{i-1} for $i = 1,\ldots,r$. The *length* of the series is the integer r.

Note that we do *not* require each G_i to be normal in G, and that a series of length r involves $r+1$ members (including G and 1).

Example 6.19. The symmetric group S_4 contains a normal subgroup V of order 4, $V = \text{gp}\{(12)(34), (13)(24)\}$, as was observed in Example 6.13; and V contains normal subgroups of order 2, for instance $U = \text{gp}\{(12)(34)\}$. So we have the following normal series for S_4, in which U is not normal in S_4:

$$S_4 \triangleright V \triangleright U \triangleright 1.$$

We could of course insert further terms, even having repetitions if we wanted.

DEFINITION. If the group G has two normal series then the second is a *refinement* of the first if each member of the first occurs as a member of the second also.

Thus a refinement of a given normal series may be made by inserting repetitions. A less trivial example is the following refinement of the above normal series for S_4:

$$S_4 \triangleright A_4 \triangleright V \triangleright U \triangleright 1.$$

DEFINITION. Two normal series for a given group G, say

$$G = G_0 \trianglerighteq G_1 \trianglerighteq \ldots \trianglerighteq G_{r-1} \trianglerighteq G_r = 1,$$

$$G = H_0 \trianglerighteq H_1 \trianglerighteq \ldots \trianglerighteq H_{s-1} \trianglerighteq H_s = 1,$$

are *isomorphic* if there is a one–one correspondence between the set of non-trivial factor groups G_{i-1}/G_i and the set of non-trivial factor groups H_{j-1}/H_j such that corresponding factor groups are isomorphic.

Example 6.20. Take G to be Z_6, the cyclic group of order 6. Two normal series for G are

$$Z_6 \rhd Z_3 \rhd 1, \quad Z_6 \rhd Z_2 \rhd 1$$

where Z_2, Z_3 denote the subgroups of orders 2, 3 respectively. We make Z_6/Z_3 correspond to $Z_2/1$, and $Z_3/1$ to Z_6/Z_2; since corresponding factor groups are isomorphic, the two series are isomorphic.

THEOREM 6.21. *Any two normal series for a given group have isomorphic refinements.*

Proof. Let the two normal series for the group G be those named in the preceding definition. We construct a refinement of the first by inserting between G_{i-1} and G_i, for each i, the following series:

$$G_{i-1} = G_i(G_{i-1} \cap H_0) \trianglerighteq G_i(G_{i-1} \cap H_1) \trianglerighteq \ldots \trianglerighteq G_i(G_{i-1} \cap H_s) = G_i.$$

Observe that $H_{j-1} \trianglerighteq H_j$ implies that

$$G_i(G_{i-1} \cap H_{j-1}) \trianglerighteq G_i(G_{i-1} \cap H_j)$$

by arguments produced in the proof of Zassenhaus's lemma (Theorem 6.18). We thus obtain a refinement of length rs. A similar refinement may be obtained for the second series by inserting between each H_{j-1} and H_j the series

$$H_{j-1} = H_j(G_0 \cap H_{j-1}) \trianglerighteq H_j(G_1 \cap H_{j-1}) \trianglerighteq \ldots \trianglerighteq H_j(G_r \cap H_{j-1}) = H_j.$$

We now claim that these refinements are isomorphic. Indeed this is a trivial consequence of Zassenhaus's lemma, which ensures at once that

$$G_i(G_{i-1} \cap H_{j-1})/G_i(G_{i-1} \cap H_j) \cong H_j(G_{i-1} \cap H_{j-1})/H_j(G_i \cap H_{j-1})$$

for $1 \leqslant i \leqslant r$ and $1 \leqslant j \leqslant s$. The theorem is now proved. □

The natural thing to do next is to take a normal series for a group G and to construct a refinement which cannot be further refined in a non-trivial manner (that is, without the insertion of repeated terms). This can certainly be done for a finite group G. If it is accomplished then the factors G_i/G_{i+1} of successive terms will have no proper normal subgroups.

DEFINITION. A *simple* group is a group in which no proper subgroup is normal.

DEFINITION. A *composition series* of a group G is a normal series

$$G = G_0 \rhd G_1 \rhd \ldots \rhd G_{r-1} \rhd G_r = 1$$

such that G_{i-1}/G_i is simple (and non-trivial), for $1 \leqslant i \leqslant r$. The groups G_{i-1}/G_i are called the *composition factors*.

The general aim of theorems about normal structure is to relate a given group to simple groups—the composition factors are a set of simple groups associated with G. A fundamental task is therefore to describe the finite simple groups. For abelian groups this is easy, for the groups in question must have no proper subgroups and are therefore of prime order (or trivial). From what we know so far these might be all the finite simple groups. In fact there are many others, and a complete classification of them is at the present time an unsolved question. Among the well-known examples are the finite alternating groups of degree five or more, which we shall study in Theorem 11.08. Meanwhile one specimen must suffice.

Example 6.22. Consider the alternating group A_5. Its non-trivial elements have order 2, e.g. (12)(34), or 3, e.g. (123), or 5, e.g. (12345), so a proper normal subgroup N contains an element of order 2 or 3 or 5. In fact there must be an element of order 3. We have

$$[(12345), (13)(24)] = (135),$$

and if N has an element of order 5 we may assume that $(12345) \in N$; and since $a \in N$ implies $[a, x] \in N$ for all $x \in A_5$ we have $(135) \in N$. Similarly if N has an element of order 2 it can be taken to be (13)(24) and again $(135) \in N$.

It follows from this that N contains all the elements in A_5 of order 3. For if $(135) \in N$ and we want to show that $(123) \in N$ then we calculate the conjugate $(235)(135)(253)$ of (135), which must lie in N; in fact it is (123), as required. A similar argument shows that if $(\alpha\beta\gamma) \in N$ then $(\alpha\beta\delta) \in N$ where α, β, γ, δ are distinct. This suffices to prove that N contains all elements of order 3 in A_5.

Every element of A_5 can be expressed as the product of an even number of transpositions, and we have, with α, β, γ, δ as above,

$$(\alpha\beta)(\alpha\gamma) = (\alpha\beta\gamma),$$
$$(\alpha\beta)(\gamma\delta) = (\alpha\beta\gamma)(\alpha\delta\gamma).$$

Thus every element of A_5 can be expressed as the product of elements of order 3 and thus lies in N. It follows that $N = A_5$ and that A_5 is simple.

Direct products of finite groups enable one to construct examples of isomorphic groups with somewhat different (though still isomorphic) composition series. The Jordan–Hölder theorem ensures that the composition series of a given group cannot differ very widely.

THEOREM 6.23 (Jordan–Hölder theorem). *In a group with a composition series every composition series is isomorphic to the given series.*

Proof. We note that a composition series of the group G cannot be properly refined to another composition series, since all the composition factors are already simple non-trivial groups (as long as $G \neq 1$, at least). Suppose two different composition series are given. Then Theorem 6.21 enables us to construct isomorphic refinements, one to each series. Since each series is isomorphic to its refinement once repeated terms are omitted, it follows that the two original series were isomorphic. This is the assertion of the theorem. □

Note that in particular the set of composition factors as well as the length of a composition series is a group invariant. But the composition factors do not determine the group, for it is easy to see that the two abelian groups of order 4 have isomorphic composition series.

COROLLARY 6.24. *Any two composition series of a finite group are isomorphic.* □

We must observe that the Jordan–Hölder theorem tells us nothing about finite abelian groups, in which all the composition series can easily be found. Neither does it tell us anything at all about the finite simple groups. Its use is with finite groups with a fair number of normal subgroups which are nevertheless non-abelian.

For infinite groups, less information on structure can be found. There are groups with no composition series in our strict sense, and even if we are prepared to liberalize our definition of composition series the Jordan–Hölder theorem may not be valid.

Example 6.25. Let G be the additive group Z of integers, and let $G_i = \mathrm{gp}\{p^i\}$, where p is an arbitrary prime, for $i = 0, 1, 2, \ldots$. Because $G_{i+1} \lhd G_i$, and G_i/G_{i+1} has order p, and $\bigcap G_i = 0$, the series

$$G = G_0 \rhd G_1 \rhd G_2 \rhd \ldots$$

has some claim to be a composition series. But the order p of G_i/G_{i+1} can be selected at will, and we can hardly claim that such a composition series yields invariants of G.

Our definitions and methods do give the following result applicable to infinite groups.

THEOREM 6.26. *Let G be a group with a composition series of length r, and let*

$$G = G_0 \rhd G_1 \rhd \ldots \rhd G_{r-1} \rhd G_r \trianglerighteq 1$$

be a normal series for G in which $G_0/G_1, \ldots, G_{r-1}/G_r$ are all simple and non-trivial. Then $G_r = 1$.

Proof. Refine the two given series according to Theorem 6.21, which states that isomorphic refinements exist. The first refinement is a composition series of length r, and therefore the second is a composition series of the same length; we omit repeated terms. It follows that $G_r = 1$, as required. □

Finite group theory is the natural setting for composition series. The idea arose from the study of polynomial equations and the resulting Galois theory, a subject which unfortunately cannot be described here. Composition series with abelian factors give rise to an extensive theory of finite soluble groups, to be touched on in Chapter 10.

The remainder of this chapter is given over to a new topic, a tool of importance in its own right but not of immediate application, which has a close affinity to normal subgroups. Up to the present we have considered operations (such as the direct product) defined on a *finite* set of groups, though infinite sets were not excluded when complications did not arise; we refer to the intersection of a set of subgroups, and the subgroup generated by a set of complexes. We shall now weave a new strand into our studies by considering products of (possibly infinite) sets of groups.

DEFINITION. The *product* of the complexes $\{C_\lambda : \lambda \in \Lambda\}$ in a given group, where the index set Λ is totally ordered,† is the union of all complexes of the form $C_{\lambda_1} C_{\lambda_2} \dots C_{\lambda_n}$, where n is a positive integer or zero, and $\lambda_1 < \lambda_2 < \dots < \lambda_n$. This product is denoted by $\prod_{\lambda \in \Lambda} C_\lambda$. When $n = 0$ the product is interpreted as the element 1.

There is a strong constraint in group theory that leads towards the above definition. In the group axioms the product $x_1 x_2$ of two elements x_1, x_2 is made meaningful, and by induction this definition may be extended to any finite number of factors. What we cannot do without special and elaborate precautions is to talk sensibly of the product of infinitely many elements. The definition of $\prod_{\lambda \in \Lambda} C_\lambda$, however, avoids this difficulty.

Example 6.27. Consider the abelian group Z_p^∞ displayed in Example 1.09. Take any prime for p, and use the additive

† See the Appendix.

notation. Define C_n to be $\{z : p^n z = 0\}$ for $n = 1, 2, \ldots$; thus Λ is the set of positive integers. Then $\sum_{n>0} C_n$ is the whole group Z_p^∞, for each element of Z_p^∞ belongs to a suitable complex C_n; if z has order p^n then $z \in C_n$.

We may in particular consider products of subgroups, even products of normal subgroups.

THEOREM 6.28. *If* $\{N_\lambda : \lambda \in \Lambda\}$ *is a non-empty set of normal subgroups of the group* G *then* $\prod_{\lambda \in \Lambda} N_\lambda$ *is also a normal subgroup.*

Proof. To prove the subgroup property we apply the usual subgroup criterion. Take $x, y \in \prod_{\lambda \in \Lambda} N_\lambda$. Then

$$x \in N_{\lambda_1} \ldots N_{\lambda_r} \quad (\lambda_1 < \ldots < \lambda_r),$$

$$y \in N_{\mu_1} \ldots N_{\mu_s} \quad (\mu_1 < \ldots < \mu_s).$$

It follows that x, y lie in $\mathrm{gp}\{N_{\lambda_1}, \ldots, N_{\lambda_r}, N_{\mu_1}, \ldots, N_{\mu_s}\}$, a subgroup which we shall call N. Therefore $xy^{-1} \in N$. But Corollary 6.10 shows that

$$N = N_{\lambda_1} \ldots N_{\lambda_r} N_{\mu_1} \ldots N_{\mu_s},$$

and $N_{\lambda_i} N_{\mu_j} = N_{\mu_j} N_{\lambda_i}$ for $1 \leqslant i \leqslant r$, $1 \leqslant j \leqslant s$. It follows that $N \subseteq \prod_{\lambda \in \Lambda} N_\lambda$, and so xy^{-1} lies in the product. So the product is a subgroup. The normal property is evident. □

We now consider the direct product and the cartesian product. The direct product of a *finite* set of groups was defined in Chapter 4, and *both* the following definitions generalize the earlier concept.

DEFINITION. The *cartesian product* of the set $\{G_\lambda : \lambda \in \Lambda\}$ of groups is the set of all functions from Λ to $\{G_\lambda\}$, so that if g is such a function then $\lambda g \in G_\lambda$; the product of two such functions f, g is defined by $\lambda(fg) = (\lambda f)(\lambda g)$. This cartesian product is denoted by $\prod_{\lambda \in \Lambda}^C G_\lambda$.

Our first task is to verify that the cartesian product G of the groups G_λ is itself a group under the given multiplication.

Closure obtains because fg is by the nature of its definition a function from Λ to $\{G_\lambda : \lambda \in \Lambda\}$. The associative law follows from that law in each factor because if f, g, $h \in G$ then

$$\lambda\{(fg)h\} = \{\lambda(fg)\}(\lambda h) = \{(\lambda f)(\lambda g)\}(\lambda h),$$

$$\lambda\{f(gh)\} = (\lambda f)\{\lambda(gh)\} = (\lambda f)\{(\lambda g)(\lambda h)\},$$

for each $\lambda \in \Lambda$. The identity element is the function f for which λf is the identity element of G_λ. The inverse f^{-1} of $f \in G$ is defined by $\lambda(f^{-1}) = (\lambda f)^{-1}$ for all $\lambda \in \Lambda$.

In the special case when Λ is a finite set $\{\lambda_1,\ldots,\lambda_n\}$ we may identify the element g of G with the following ordered set of n elements:

$$(\lambda_1 g,\ldots,\lambda_n g).$$

It follows that the cartesian product G is simply the direct product $G_{\lambda_1} \times \ldots \times G_{\lambda_n}$ as defined in Chapter 4.

Definition. If g is a function from a set Λ to the set $\{G_\lambda : \lambda \in \Lambda\}$ of groups, then the *support* of g is $\{\mu : \mu g \neq 1\}$.

Definition. The *direct product* of the set $\{G_\lambda : \lambda \in \Lambda\}$ of groups is the set of all functions from Λ to $\{G_\lambda : \lambda \subset \Lambda\}$, each having finite support; the product of two such functions f, g is defined by $\lambda(fg) = (\lambda f)(\lambda g)$. This direct product is denoted by $\prod_{\lambda \in \Lambda}^{D} G_\lambda$.

It is easy to prove that the direct product is also a group. The key point is the fact that if σ_1, σ_2 denote the supports of $f_1, f_2 \in \prod_{\lambda \in \Lambda}^{D} G_\lambda$ then the support of $f_1 f_2$ lies in $\sigma_1 \cup \sigma_2$, and the support of f_1^{-1} is σ_1. Of course, the identity function (which has empty support) is assigned finite support by courtesy.

It is also easy to see that the cartesian product $\prod_{\lambda \in \Lambda}^{C} G_\lambda$ contains a subgroup isomorphic to $\prod_{\lambda \in \Lambda}^{D} G_\lambda$; and that if Λ is finite then the new definition of direct product coincides with the earlier.

THEOREM 6.29. *The direct product G of the groups $\{G_\lambda : \lambda \in \Lambda\}$ contains subgroups H_λ which are isomorphic to G_λ for each λ and which have the following properties:*

(i) *H_λ commutes with H_μ element by element if $\lambda \neq \mu$;*

(ii) $G = \prod_{\lambda \in \Lambda} H_\lambda$;

(iii) $H_\lambda \cap \prod_{\mu \neq \lambda} H_\mu = 1$ *for each* $\lambda \in \Lambda$.

Proof. We define H_λ as the set of functions h such that $\mu h = 1$ for all μ except λ. Obviously H_λ is a subgroup isomorphic to G_λ.

(i) Take $h_\lambda \in H_\lambda$, $h_\mu \in H_\mu$, with $\lambda \neq \mu$; and take an arbitrary element ν in Λ. Then

$$\nu(h_\lambda h_\mu) = (\nu h_\lambda)(\nu h_\mu),$$
$$\nu(h_\mu h_\lambda) = (\nu h_\mu)(\nu h_\lambda).$$

If ν is neither λ nor μ then both right-hand sides are trivial. If $\nu = \lambda$ then both equal νh_λ, and if $\nu = \mu$ then both equal νh_μ. In any case $\nu(h_\lambda h_\mu) = \nu(h_\mu h_\lambda)$ and $h_\lambda h_\mu = h_\mu h_\lambda$, as required.

(ii) Let g be an arbitrary element of G. Then

$$\lambda g = 1 \quad \text{if } \lambda \notin \{\lambda_1, \dots, \lambda_n\}$$

for some $n \geqslant 0$. We define h_i for $1 \leqslant i \leqslant n$ by

$$\lambda_i h_i = \lambda_i g, \quad \lambda h_i = 1 \text{ for } \lambda \neq \lambda_i.$$

Then $g = h_1 \dots h_n$. It follows that $G = \prod_{\lambda \in \Lambda} H_\lambda$, as claimed.

(iii) $\prod_{\mu \neq \lambda} H_\mu$ contains just those functions h of G such that $\lambda h = 1$; this follows from the definition of product. The only such function h in H_λ is the identity function. □

Note carefully that Λ need not be a totally ordered set for the product in (ii) to be meaningful, as (i) implies that the order of factors is immaterial.

The above theorem is concerned with a situation in which the direct product has been *constructed* from a given set of groups. The next is a companion theorem in that it gives a criterion for a *given* group to be isomorphic to the direct product of certain subgroups.

THEOREM 6.30. *Let the group G have subgroups $\{G_\lambda : \lambda \in \Lambda\}$ satisfying the conditions*:

(i) *G_λ commutes with G_μ element by element if $\lambda \neq \mu$;*

(ii) *$G = \prod_{\lambda \in \Lambda} G_\lambda$ (note again that a total ordering on Λ is not needed);*

(iii) *$G_\lambda \cap \prod_{\mu \neq \lambda} G_\mu = 1$ for each $\lambda \in \Lambda$.*

Then $G \cong \prod_{\lambda \in \Lambda}^{D} G_\lambda$.

Proof. Take an arbitrary element g of G. By (ii) it may be written as $g_{\lambda_1} \dots g_{\lambda_n}$ with $g_{\lambda_i} \in G_{\lambda_i}$ for $1 \leqslant i \leqslant n$ and $\lambda_i \in \Lambda$. Indeed g has a unique representation of this form, for if it had two then use of (i) and (iii) would show that corresponding factors are the same.

We may now construct a mapping ϕ from G to $\prod_{\lambda \in \Lambda}^{D} G_\lambda$. We define $g\phi$ to be that element of the direct product which has support $\{\lambda_1, \dots, \lambda_n\}$ and which maps λ_i onto g_{λ_i} for $1 \leqslant i \leqslant n$; this is properly defined as g determines g_{λ_i} uniquely.

Suppose that h lies in G, so that h also has a unique representation as the product of factors from a finite number of G_λ. It is clear that matters regarding g and h may be so arranged that

$$g = g_{\lambda_1} \dots g_{\lambda_m}, \qquad h = h_{\lambda_1} \dots h_{\lambda_m}$$

where $g_{\lambda_i}, h_{\lambda_i} \in G_{\lambda_i}$ for $1 \leqslant i \leqslant m$, provided that some g_{λ_i} and h_{λ_i} are allowed to be 1. Then the supports of $g\phi$ and $h\phi$ will be subsets of $\{\lambda_1, \dots, \lambda_m\}$, and

$$\lambda_i\{(g\phi)(h\phi)\} = \{\lambda_i(g\phi)\}\{\lambda_i(h\phi)\} = g_{\lambda_i} h_{\lambda_i}$$

for $1 \leqslant i \leqslant m$. On the other hand $gh = (g_{\lambda_1} h_{\lambda_1}) \dots (g_{\lambda_m} h_{\lambda_m})$, so that the support of $(gh)\phi$ lies in $\{\lambda_1, \dots, \lambda_m\}$, and

$$\lambda_i\{(gh)\phi\} = g_{\lambda_i} h_{\lambda_i}$$

for $1 \leqslant i \leqslant m$. This shows that $(g\phi)(h\phi) = (gh)\phi$, and so ϕ is a homomorphism from G to $\prod_{\lambda \in \Lambda}^{D} G_\lambda$.

But ϕ is one–one. If $g\phi = h\phi$ then $\lambda(g\phi) = \lambda(h\phi)$ for all $\lambda \in \Lambda$, and it follows that the product representations of g and h coincide factor by factor, so that $g = h$. Further ϕ is onto the direct product; given the function g in $\prod_{\lambda \in \Lambda}^{D} G_\lambda$, with support $\{\lambda_1,\ldots,\lambda_m\}$ such that $\lambda_i g = g_{\lambda_i}$ for $1 \leqslant i \leqslant m$, we see that it is the image under ϕ of $g_{\lambda_1} \ldots g_{\lambda_m} \in G$.

The required isomorphism $G \cong \prod_{\lambda \in \Lambda}^{D} G_\lambda$ has now been established. □

We remark that there is a closely related theorem in which (i) is replaced by: $G_\lambda \trianglelefteq G$ for each λ.

All our examples of finite direct products are also examples of general cartesian and direct products. A less trivial illustration follows.

Example 6.31. Let A be a periodic abelian group. There is clearly in A a subgroup A_p consisting of all elements of p-power order, for each prime p. We shall prove by means of the above theorem that A is, or more strictly is isomorphic to, the direct sum of all these subgroups. (As usual we speak of the direct sum and the cartesian sum when abelian groups are involved.) There is no trouble over the commutative property (i) or property (iii). Regarding (ii), we have seen (in the proof of Theorem 5.01) that if A_0 is a finite abelian group then an element a of order $p_1^{\alpha_1} \ldots p_r^{\alpha_r}$, where $p_1,\ldots, p_r$ are distinct primes, is the sum of elements of orders $p_1^{\alpha_1},\ldots, p_r^{\alpha_r}$. The same holds good in A for if $a \in A$ then $\mathrm{gp}\{a\}$ is finite and can be taken to be A_0. Thus $a \in \sum_{i=1}^{r} A_{p_i}$, and it follows that $A = \sum A_p$, the sum being taken over all primes p. Hence $A = \sum^D A_p$.

It is rather more difficult to handle the cartesian product than the direct product, and so it is fortunate that our main use for the latter will be to construct examples. It is certainly true that $\prod_{\lambda \in \Lambda}^{C} G_\lambda$ contains copies of each G_λ, for these may be found in its subgroup $\prod_{\lambda \in \Lambda}^{D} G_\lambda$. We have properties (i) and (iii) of the direct product given in the above theorems, for the same reason, but (ii) is by no means valid. In other words, $\{g_\lambda : \lambda g_\lambda \in G_\lambda, \mu g_\lambda = 1 \text{ for } \mu \neq \lambda\}$ is not a set of generators for $\prod^C_{\lambda \in \Lambda} G_\lambda$ when Λ is infinite, for no element of infinite support lies in the subgroup it generates, which is precisely $\prod^D_{\lambda \in \Lambda} G_\lambda$.

We shall, however, prove one elementary theorem about cartesian products.

THEOREM 6.32. *If $\{N_\lambda : \lambda \in \Lambda\}$ is a set of normal subgroups of the group G and if $N = \bigcap_{\lambda \in \Lambda} N_\lambda$ then G/N is isomorphic to a subgroup of $\prod^C_{\lambda \in \Lambda} (G/N_\lambda)$.*

Proof. We shall define a mapping ϕ from G into the cartesian product $P = \prod^C_{\lambda \in \Lambda} G/N_\lambda$, and apply the first isomorphism theorem. Take an arbitrary element x in G.

Let $x\phi$ be the function $g \in P$ such that $\lambda g = N_\lambda x$ for each λ in Λ. To prove that ϕ is a homomorphism, suppose $y \in G$ maps onto $h \in P$ so that $\lambda h = N_\lambda y$. Then $(xy)\phi$ is the function $f \in P$ for which $\lambda f = N_\lambda xy$, and we have

$$\lambda f = N_\lambda xy = (N_\lambda x)(N_\lambda y) = (\lambda g)(\lambda h) = \lambda(gh)$$

for all $\lambda \in \Lambda$. Therefore $f = gh$, that is

$$(xy)\phi = (x\phi)(y\phi).$$

and ϕ is a homomorphism.

Let K be the kernel of ϕ. If $x \in K$ then $x\phi$ is the identity function, so that $N_\lambda x = N_\lambda$ for all $\lambda \in \Lambda$. This means precisely that $x \in N_\lambda$, and it follows that $x \in N$. Therefore $K \leqslant N$. Conversely, it is clear that if $x \in N$ then $x \in N_\lambda$ and $x\phi$ is the identity function, so that $N \leqslant K$. It follows that the kernel of ϕ is N.

By the first isomorphism theorem, then, G/N is isomorphic to the image of G in P under ϕ, and this is clearly a subgroup of P.□

Example 6.33. Let G be the additive group Z of integers, and let $N_n = \text{gp}\{n\}$ for $n = 1, 2, \ldots$. Then $N = \bigcap_{n>0} N_n = 0$, so that $G/N \cong Z$. On the other hand, G/N_n is isomorphic to the additive group of residues modulo n. The theorem asserts that $\prod^C_{n>0} G/N_n$ contains an infinite cyclic subgroup. Indeed the proof specifies a generator for this subgroup, namely the function g such that $ng = N_n+1$.

Note, however, that $\prod^D_{n>0} G/N_n$ does not contain an infinite cyclic subgroup because it is a periodic group.

Problems

1. Find a composition series for the subgroup of the symmetric group S_7 generated by $\{(1234567), (243756)\}$. Find a second in which the first composition factor has a different order.

2. Find a normal series of arbitrary positive length n for Z_p^∞ (see Example 1.09) such that $n-1$ of its factor groups have order p. Deduce that Z_p^∞ has no composition series.

3. Prove that the only finite class of conjugate elements in an infinite simple group is $\{1\}$.

4. The group G contains subgroups A, N_1, N_2 such that $N_1 \leqslant N_2$, $N_1 \trianglelefteq G$, $N_2 \trianglelefteq G$. Prove that $(A \cap N_2)/(A \cap N_1)$ is isomorphic to a subgroup of N_2/N_1.

Illustrate this result in the case when $G = S_4$, $N_1 = V$, $N_2 = A_4$, $A = \text{gp}\{(12), (123)\}$, in the notation of the text.

5. A group G is said to be *metacyclic* if it has a normal subgroup N such that both N and G/N are cyclic. Prove that every subgroup and every factor group of a metacyclic group are again metacyclic.

*6. (i) Prove that if the group $G = AB$ where $A \leqslant G$, $B \leqslant G$, then there is a one–one correspondence between the cosets of B in AB and the cosets of $A \cap B$ in A.

(ii) Prove that if G is a group with subgroups A, B of (finite) coprime indices then $G = AB$.

7. Prove that a finite group of order p^n has a composition series of length n.

8. The subgroup T_n of the unrestricted symmetric group G on the positive integers is generated by the transposition $(2n-1, 2n)$ for $n = 1, 2, \ldots$. Show that $\text{gp}\{T_n : n > 0\}$ is $\prod^D_{n>0} T_n$.

Prove that G also contains a subgroup isomorphic to $\prod^C_{n>0} T_n$.

9. Let N be a normal subgroup of the group G and let A be an arbitrary subgroup of G containing N. Prove that there is a one–one correspondence between the cosets of A in G and the cosets of A/N in G/N. What relation does this result bear to the third isomorphism theorem?

10. Let G_1, G_2 be isomorphic groups with normal subgroups N_1, N_2 respectively. Prove that if the isomorphism between G_1 and G_2 induces (by restriction) an isomorphism between N_1 and N_2 then $G_1/N_1 \cong G_2/N_2$.

11. Prove the Jordan–Hölder theorem for finite groups by using induction on the least integer for which there exists a composition series of that length and by appealing to the second isomorphism theorem.

12. Let N_1, N_2 be normal subgroups of G, and let N be a subgroup of G which is a normal subgroup of both N_1 and N_2. Prove that

$$(N_1/N)(N_2/N) = (N_1 N_2)/N.$$

Let A be a subgroup of G and let M and N be normal subgroups of G with $M \geqslant N$. Show that the image of A, when G is mapped onto G/N and G/N onto $(G/N)/(M/N)$ by natural homomorphisms, $\cong MA/M$.

13. Lemma 6.14 in the text shows that if ϕ is a homomorphism from G_1 into G_2 mapping N_1 into N_2 (where N_i is a normal subgroup of G_i for $i = 1, 2$) then there is a homomorphism ψ from G_1/N_1 into G_2/N_2. Suppose that ϕ is onto G_2 and its restriction to N_1 is onto N_2. Prove then that ψ is an isomorphism if and only if the kernel of ϕ lies in N_1.

14. Let A, B, C be subgroups of the group G. Prove that ABC is a subgroup if and only if $ABC = BCA$ and $BCB \subseteq ABC$.

Let G be the multiplicative group of all matrices of the form

$$\begin{pmatrix} 1 & a & c \\ 0 & 1 & b \\ 0 & 0 & 1 \end{pmatrix},$$

where a, b, c are integers. Take A, B, C to be the subgroups generated by these matrices, respectively:

$$\begin{pmatrix} 1 & 1 & 0 \\ 0 & 1 & 0 \\ 0 & 0 & 1 \end{pmatrix}, \quad \begin{pmatrix} 1 & 0 & 0 \\ 0 & 1 & 1 \\ 0 & 0 & 1 \end{pmatrix}, \quad \begin{pmatrix} 1 & 0 & 1 \\ 0 & 1 & 0 \\ 0 & 0 & 1 \end{pmatrix}.$$

Prove that $G = ABC$. Is AB a subgroup of G?

15. Let the group G have, as a subgroup of index 2, the simple non-abelian group H. Prove that H is the unique subgroup of index 2, and deduce that it is characteristic in G.

Deduce that S_5 has only one proper normal subgroup.

Is a non-abelian subgroup of index two always characteristic?

*16. Let C and D denote respectively the cartesian and the direct sum of groups of order p, as p takes every prime value. Prove that if a is any element of C/D then the equation $nx = a$ has a solution x for any positive integer n. (Additive notation is used here.)

**17. Prove that the dihedral group D_n of order 2^n has the property that $D_n/\zeta(D_n) \cong D_{n-1}$ for $n \geqslant 3$ (D_n is described in Example 1.25, and D_2 is to be taken as $C_2 \times C_2$ where C_2 is the group of order 2).

Deduce that if $D = \prod_{n \geqslant 2}^{D} D_n$ then $\zeta(D) \neq 1$ while $D/\zeta(D)$ is isomorphic to D.

**18. The infinite dihedral group D is generated by elements a and b, each of order 2, with ab of infinite order (see Problem 5 of Chapter 3). Prove that $N_n = \mathrm{gp}\{(ab)^{2^n}\}$ is a normal subgroup of D for $n = 1, 2, \ldots$.

Let D_n denote the factor group D/N_n. Show that D_n is a group of order 2^{n+1} and is in fact the finite dihedral group of that order (see Example 1.25 and the preceding problem).

Deduce that $\prod_{n>0}^{C} D_n$ has a subgroup isomorphic to D.

7

Sylow theorems

THREE remarkable classical theorems of Sylow deal with the existence of certain subgroups in a finite group and with their properties in relation to the group. We shall focus attention on questions of existence, to start with. Lagrange's theorem 3.15 states that if G is a finite group, then $|G|$ is divisible by the order of every subgroup. But this tells us nothing about the *existence* in G of subgroups having for their orders prescribed factors of $|G|$.

In fact subgroups with such arbitrary orders may not exist, an assertion that was made in Chapter 3 and that will now be proved.

Example 7.01. The alternating group A_4 of order 12 has no subgroup of order 6. For suppose that there were a subgroup S with six elements. Since A_4 contains three elements of order 2 and eight of order 3, there must be at least two elements of order 3 in S. If S contains three elements of order 2 and just two of order 3, we see that S contains the normal subgroup of A_4 that has order 4 and so $S = A_4$, a contradiction. By appropriately naming the set on which A_4 acts, we may assume that S contains (123) and (124). But now a little calculation produces seven distinct elements in S:

$$1, (123), (132), (124), (142), (123)(124) = (14)(23),$$
$$(124)(123) = (13)(24).$$

This contradiction shows that S does not exist.

Sylow's first theorem gives a condition for existence of subgroups. First we prove:

THEOREM 7.02. *Let G be a group of order $p^\alpha r$ where p is prime, $\alpha > 0$, and r is not divisible by p. Then G contains a subgroup of order p^β for each β satisfying $0 \leqslant \beta \leqslant \alpha$.*

Proof. We use induction on the order of G. The inductive process starts with a group G in which $\alpha = 1$ and $r = 1$; it is clear that G has order p, as does any group satisfying our hypotheses and having no proper subgroup.

In general, we assume that every group with smaller order than G has p-subgroups of the kind required. In particular, this will be true of all proper subgroups of G. Suppose there is a proper subgroup S of index prime to p. It follows that p^α is the highest power of p to divide $|S|$. By the inductive hypothesis, S has subgroups of order p^β for each β in $0 \leqslant \beta \leqslant \alpha$; since these are also subgroups of G, this case is finished.

We are now left with the case in which the index of every proper subgroup of G is divisible by p. Lemma 4.28 is now applicable. Thus $\zeta(G)$ is an abelian subgroup of G, with order m divisible by p; and it must contain an element of order p. This is clear from Sylow's first theorem for abelian groups—we refer to Corollary 5.03. Therefore $\zeta(G)$ contains a subgroup N of order p. This is normal in G as it is central in G.
o' Thus $\zeta(G)$ is an abelian subgroup of G, with order m divisible by p; and it must contain an element of order p. This is clear from Sylow's first theorem for abelian groups—we refer to Corollary 5.03. Therefore $\zeta(G)$ contains a subgroup N of order p. This is normal in G as it is central in G.

We next consider G/N and apply the inductive hypothesis, as we may because G/N has order $p^{\alpha-1}r$. We conclude that if $0 < \beta \leqslant \alpha$ then G/N has a subgroup, M/N say, of order $p^{\beta-1}$ for each such β. Note the use of Theorem 6.04 here. It is now clear that M has order p^β, and the proof is finished by the remark that the case $\beta = 0$ is trivial. □

COROLLARY 7.03 (Sylow's first theorem). *If the group G has order $p^\alpha r$, where the prime p does not divide r, then G has at least one subgroup of order p^α.* □

COROLLARY 7.04. *If the prime p divides the order of the finite group G, then G contains at least one element of order p.*

Proof. Take $\beta = 1$ in the theorem; the generator of a subgroup of order p itself has order p. □

COROLLARY 7.05. *The set of primes dividing the order of a finite group is the set of primes occurring as orders of elements in it.*

Proof. Use Corollary 7.04 and Lagrange's theorem. □

COROLLARY 7.06. *A finite group is a p-group if and only if it has p-power order.* □

(We note for the sake of completeness that a finite group G satisfying the hypotheses of Lemma 4.29 must have p-power order. For if a prime distinct from p divides $|G|$, then there is a proper subgroup of p-power order whose index is not divisible by p.)

The subgroups of order p^α which appeared in Corollary 7.03 play a fundamental role in the theory of finite groups. It is on their account that we introduce new terminology. Let π be a set of primes (which we shall assume contains at least one prime in order to avoid hair-splitting). We say that a positive integer is a *π-number* if all its prime factors lie in π.

DEFINITION. A *π-group* is a group in which the order of every element is a π-number.

Thus Corollary 7.05 makes it plain that a finite group is a π-group if and only if its order is a π-number. We use the term 'π-subgroup' in the obvious sense.

DEFINITION. A *Sylow π-subgroup* of a group G is a π-subgroup which is not properly contained in any π-subgroup of G.

Thus Corollary 7.03 asserts that in the finite group G there is a Sylow p-subgroup of order p^α; note that there can be no subgroup of strictly larger p-power order, by Lagrange's theorem.

A finite group G has Sylow π-subgroups for every choice of π. Indeed, it is easily shown that every π-subgroup lies in a Sylow π-subgroup. To start with G has π-subgroups, for 1 is a π-subgroup. And if S is a π-subgroup which is not a Sylow π-subgroup then S is properly contained in another π-subgroup. After a finite number of steps this process must lead to a Sylow π-subgroup, because G is finite.

This argument does not prove Sylow's first theorem when $\pi = \{p\}$, of course, for it does not prove that the group of order $p^{\alpha}r$ has a subgroup of order p^{α}.

A natural question to ask at this point is whether Sylow's first theorem can be generalized in the following way. Let G be a finite group, and let m be the largest π-number dividing $|G|$, for some set of primes π. When $\pi = \{p\}$ we now know that G has a subgroup of order m, and one might ask if something similar can be proved for general π. That this is not so without restrictions is shown by the next example, and what the restrictions are will be discussed in Chapter 10. Notice, by the way, that Example 7.01 is irrelevant to the present discussion, because 6 is not the largest π-number dividing 12 when $\pi = \{2, 3\}$.

Example 7.07. Consider the π-subgroups of A_5, where $\pi = \{2, 5\}$. We first show that A_5 has no subgroup of order 20 (which is the maximal π-number dividing $|A_5|$). For if there were such a subgroup of index 3, then it would have at most three conjugates; the intersection of these would be a normal subgroup of index at most $3^3 = 27$, as the method of Theorem 4.16 shows. But this would be a proper normal subgroup, whereas A_5 is simple according to Example 6.22. It follows that A_5 has no subgroup of order 20.

Nevertheless, A_5 must have Sylow π-subgroups. What are they? Any Sylow 2-subgroup, such as gp{(12)(34), (13)(24)}, is one; if it were contained in a larger π-subgroup then the latter would have order 20 by Lagrange's theorem. Another Sylow π-subgroup is gp{(12345), (25)(34)}; for it will be found that this has order 10 and is therefore again a maximal π-subgroup. Here is a case, then, in which Sylow π-subgroups have different orders.

We can proceed in several ways. This state of affairs arises because we wish to emphasize both the elegance of the classical Sylow theorems and the difficulties that arise when we try to generalize them. What we shall do is to state Sylow's second and third theorems at once, giving some account of their applications as well, and to prove them in the course of a careful analysis of π-subgroups of general (perhaps infinite) groups. The reader who wishes for nothing but proof of Sylow's theorems will be able to select what he wants from the results to follow. Alternative proofs to all three classical theorems will be found in an Appendix to this chapter.

SYLOW'S SECOND THEOREM.† *In a finite group the Sylow p-subgroups (for a fixed prime p) are all conjugate and therefore isomorphic.*

SYLOW'S THIRD THEOREM. *In a finite group the number of Sylow p-subgroups (for a fixed prime p) is congruent to* 1 *modulo p.*

Example 7.08. Consider the Sylow 2-subgroups of A_5. By Sylow's first theorem they each have order 4. One of them is $G_2 = \text{gp}\{(12)(34), (13)(24), (14)(23)\}$. Now A_5 has a subgroup isomorphic in a natural way to the alternating group on $\{1, 2, 3, 4\}$ and G_2 is normal in this (compare Example 6.13). Therefore $|A_5 : N(G_2)|$ is 1 or 5. In fact this index is 5, because A_5 is simple (Example 6.22). Take a right transversal of $N(G_2)$ in A_5; one might choose $\{a^i : 0 \leqslant i < 5\}$ where $a = (12345)$, for instance. Then the conjugacy class of G_2 is $\{G_2^{a^i} : 0 \leqslant i < 5\}$, by Theorem 4.12. Sylow's second theorem allows us to assert that this conjugacy class is the set of all Sylow 2-subgroups of A_5. Note that there are 5 of them, and $5 \equiv 1$ modulo 2 in accordance with the third theorem.

Example 7.09. Let G be any group of order 40; we prove that G must contain a normal subgroup of order 5.

The first Sylow theorem implies that G has a subgroup of order 5. By the third theorem there are in fact $1+5k$, for some integer k. By the second theorem $1+5k$ divides 40, because the Sylow 5-subgroups are all conjugate and so the normalizer of any has index $1+5k$ (recall Corollary 4.14). But $1+5k$ divides 40 if and only if $k = 0$, so there is just one Sylow 5-subgroup. Since it coincides with all its conjugates, it is a normal subgroup of G.

† A slight liberty has been taken in the numbering of Sylow's three theorems.

Example 7.10. Next we prove that an arbitrary group G of order 12 has a normal subgroup of order 3 or 4. Sylow's theorems, used as in the previous example, show that the number of subgroups of order 3 is either one or four. Suppose that there are four. Each contains two elements of order 3, by Lagrange's theorem. We thus find eight elements of order 3, which are easily seen to be distinct. There are four other elements of G. By Sylow's first theorem there is at least one subgroup of order 4. In the present circumstances there is precisely one, and by the second theorem it is normal. Therefore there is a normal subgroup, if not of order 3 then of order 4.

We now embark on an attempt to chart some of the territory of Sylow properties, exploring such parts of it as we can reach. Some of our theorems will be mere formal implications to start with, demanding conditions but not specifying when these are satisfied: this is simply part of our analysis. One difficulty which presents itself at once is this: how do we know that Sylow π-subgroups exist in infinite groups? This is a tricky set-theoretical point. But we shall make no assumption concerning it and therefore need not even explain what the difficulties are—the hypotheses of our theorems will always guarantee the existence of Sylow π-subgroups, except for finite groups, since we know that they exist in that case.

THEOREM 7.11. *If G_π is a Sylow π-subgroup of the group G, and if α is an automorphism of G, then $G_\pi \alpha$ is a Sylow π-subgroup of G.*

Proof. Suppose by way of contradiction that $G_\pi \alpha$ is not a Sylow π-subgroup of G. It is clear that $G_\pi \alpha$ is a π-subgroup. Therefore $G_\pi \alpha$ is properly contained in some π-subgroup, G'_π say. Then $(G'_\pi)\alpha^{-1}$ is a π-subgroup because α^{-1} is an automorphism of G; and $(G'_\pi)\alpha^{-1}$ contains G_π. But this inclusion is proper, because if $x \in G'_\pi$ and $x \notin G_\pi \alpha$ then $x\alpha^{-1} \in (G'_\pi)\alpha^{-1}$ and $x\alpha^{-1} \notin G_\pi$. The fact that G_π is properly contained in a π-subgroup is a contradiction as G_π was supposed to be a Sylow π-subgroup. □

COROLLARY 7.12. *Any conjugate of a Sylow π-subgroup of G is also a Sylow π-subgroup of G.*

Proof. Use Theorem 4.32. □

One of the general problems of Sylow theory is to find reasonable conditions under which the obvious converse of Corollary 7.12 (the Sylow π-subgroups all lie in one conjugacy class of subgroups) holds. A similar question emerges from the form of the next theorem, which is usually known as Frattini's lemma.

THEOREM 7.13. *Let the group G have a normal subgroup K with the following properties:*

(i) *a Sylow π-subgroup of K exists;*
(ii) *if K_π, K'_π are two Sylow π-subgroups of K such that $K'_\pi = K_\pi \alpha$ for some automorphism α of K, then K_π and K'_π are conjugate in K.*

Then $G = KN(K_\pi)$ where K_π is any Sylow π-subgroup of K.

Proof. As usual $N(K_\pi)$ denotes the normalizer of K_π in G.

Take any element x in G. Since $K \trianglelefteq G$ and $K_\pi \leqslant K$, we have $K_\pi^x \leqslant K$. Now conjugation by x induces an automorphism of K, by Theorem 4.32, and K_π^x corresponds to K_π in this automorphism. By Theorem 7.11, therefore, K_π^x is a Sylow π-subgroup of K. By hypothesis, there exists k in K such that $K_\pi^x = K_\pi^k$. But this implies that $kx^{-1} \in N(K_\pi)$, and so $x^{-1} \in KN(K_\pi)$. This proves Theorem 7.13. □

COROLLARY 7.14. *Let the group G have a Sylow π-subgroup G_π and a subgroup K containing $N(G_\pi)$ and having property* (ii) *of the above theorem. Then $K = N(K)$.*

Proof. By $N(K)$ we mean the normalizer of K in G.

The normalizers of G_π in G and $N(K)$ are $N(G_\pi)$ and $N(K) \cap N(G_\pi)$, respectively. Since K is normal in $N(K)$, by definition of normalizer, we may apply Theorem 7.13 to the group $N(K)$. The conclusion is that $N(K) = K(N(K) \cap N(G_\pi))$. But $N(G_\pi) \leqslant K$ by hypothesis. It follows that $N(K) = K$. □

The next theorem does not mention Sylow π-subgroups, but it has no less than six corollaries which are of importance in our work.

THEOREM 7.15. *If A and B are π-subgroups of the group G, with A normal in G, then AB is a π-subgroup of G.*

Proof. Corollary 6.10 ensures that AB is a subgroup. By the second isomorphism theorem, $AB/A \cong B/(A \cap B)$. Now $B/(A \cap B)$ is a π-group because B is a π-group. Therefore, if $x \in AB$ then $x^n \in A$ for some positive π-number n (depending on x). But the order of x^n is a π-number since A is a π-group. It follows that the order of x is a π-number, and so AB is a π-group. □

COROLLARY 7.16. *A normal π-subgroup of the group G lies in every Sylow π-subgroup of G.*

Proof. Let A be a normal π-subgroup and B be any Sylow π-subgroup. Then AB is a π-subgroup by the theorem. But clearly AB contains the Sylow π-subgroup B. It follows that $AB = B$, and so $A \leqslant B$, as required. □

COROLLARY 7.17. *A group has precisely one Sylow π-subgroup if and only if it has a normal Sylow π-subgroup.*

Proof. Suppose G has precisely one Sylow π-subgroup. By Corollary 7.12 this must coincide with all its conjugates; in other words, it is normal.

Suppose G has a normal Sylow π-subgroup. By Corollary 7.16 this lies in, and therefore coincides with, every Sylow π-subgroup. Thus G has a unique Sylow π-subgroup. □

COROLLARY 7.18. *A group has precisely one Sylow π-subgroup if and only if it has a characteristic Sylow π-subgroup.*

Proof. A characteristic subgroup is normal, and this remark with the previous corollary disposes of one half of the proof.

If the group G has precisely one Sylow π-subgroup, G_π say, and if α is any automorphism of G, then $G_\pi \alpha$ is also a Sylow π-subgroup of G according to Theorem 7.11, and $G_\pi \alpha = G_\pi$. That is to say, G_π is a characteristic subgroup of G. □

COROLLARY 7.19. *If G_π is a Sylow π-subgroup of G then G_π is the unique Sylow π-subgroup of $N(G_\pi)$.*

Proof. Use Corollary 7.17 in the group $N(G_\pi)$. □

COROLLARY 7.20. *If G_π is a Sylow π-subgroup of G then no distinct conjugate of G_π lies wholly in $N(G_\pi)$.*

Proof. This follows from Corollaries 7.12 and 7.19. □

COROLLARY 7.21. *The normalizer of a Sylow π-subgroup of G is its own normalizer.*

Proof. Let $x \in N(N(G_\pi))$ where G_π is a Sylow π-subgroup of G. This means that $N(G_\pi)^x \leqslant N(G_\pi)$. But since $G_\pi \leqslant N(G_\pi)$ we also have $G_\pi^x \leqslant N(G_\pi)$. By Corollary 7.12 G_π^x is a Sylow π-subgroup of G, and we deduce from Corollary 7.19 that $G_\pi^x = G_\pi$. That is to say, $x \in N(G_\pi)$. We have shown that $N(N(G_\pi)) \leqslant N(G_\pi)$, establishing the desired result. □

Note the relation between Corollaries 7.14 and 7.21.

We now move towards incisive results. What has been said above should make it plain that very special assumptions will be needed. Accordingly our results deal with the case in which $\pi = \{p\}$, and we shall resume the discussion of general π-subgroups of finite groups in Chapter 10. One further lemma only is general in nature.

LEMMA 7.22. *The index of any subgroup in a π-group is either a π-number or infinite.*

Proof. Let S be a subgroup of finite index in the π-group G. Theorem 4.16 guarantees the existence of a subgroup N of finite index in G such that $N \leqslant S$ and $N \trianglelefteq G$. It is clear that G/N is a finite π-group, and so by Corollary 7.05 the order of G/N is a π-number. Therefore Lagrange's theorem gives the fact that $|G/N : S/N|$ is a π-number.

What is next required is a proof that there is a one–one correspondence between the right cosets of S/N in G/N and the right cosets of S in G. Now the mapping that sends the coset

Sx onto the coset $(S/N)Nx$ is clearly onto, and we claim that it is one–one. Suppose then that $(S/N)Nx = (S/N)Ny$. It follows that $(Nx)(Ny)^{-1} \in S/N$, and so $xy^{-1} \in S$; therefore $Sx = Sy$. We have proved that the mapping is one–one, establishing the desired correspondence.

Thus the index of S in G is a π-number. □

THEOREM 7.23. *Let G be a group with a Sylow p-subgroup G_p which has only a finite number of conjugates in G. Then*

(i) *G has $1+kp$ Sylow p-subgroups, for some integer k;*

(ii) *every Sylow p-subgroup of G is conjugate to G_p.*

Proof. We define certain sets of conjugates of G_p in the following manner. Let Q be any p-subgroup of G, and put

$$\mathscr{P}(Q) = \{X : X \text{ is conjugate to } G_p, \text{ and } Q \nleqslant N(X)\}.$$

Thus every $\mathscr{P}(Q)$ is a finite set; it may be empty, and if $G_p = 1$ it will be empty. By Corollary 7.12 the elements of $\mathscr{P}(Q)$ are Sylow p-subgroups of G.

Now take $X \in \mathscr{P}(Q)$. We wish to prove that if $q \in Q$ then $X^q \in \mathscr{P}(Q)$. For if $X^q \notin \mathscr{P}(Q)$ then Q normalizes X^q (by definition of $\mathscr{P}(Q)$). It is an easy step to deduce that Q normalizes X itself, but this is impossible because $X \in \mathscr{P}(Q)$ and so Q cannot normalize X. Therefore $X^q \in \mathscr{P}(Q)$.

Next we consider the number of elements in $\{X^q : q \in Q\}$. It is finite because G_p has only finitely many conjugates in G. But this number equals $|Q : Q \cap N(X)|$ by Corollary 4.14, and $Q \cap N(X)$ has p-power index in Q by Lemma 7.22. As Q does not normalize X (by definition of $\mathscr{P}(Q)$) this index cannot be 1. Thus $\{X^q : q \in Q\}$ has p^n elements for some $n > 0$.

As we take different elements X in $\mathscr{P}(Q)$ the resulting sets $\{X^q : q \in Q\}$ clearly form a partition of $\mathscr{P}(Q)$. The conclusion of the last paragraph shows that $\mathscr{P}(Q)$ has kp elements in all, for some $k \geqslant 0$. We shall apply this fact with two special choices of Q.

Take Q to be G_p. Now G_p is the only conjugate of G_p in G lying in $N(G_p)$, by Corollary 7.19. It follows that $\mathscr{P}(G_p)$ contains all the conjugates of G_p except G_p itself. Since $\mathscr{P}(G_p)$ has kp elements, for some $k \geqslant 0$, G_p has $1+kp$ conjugates.

Now let Q be an arbitrary Sylow p-subgroup, G'_p say, of G. We know (from above) that $\mathscr{P}(G'_p)$ has $k'p$ members for some k', and these are all conjugates of G_p. But they cannot be all the conjugates of G_p, for these number $1+kp$. (We can never have $k'p = 1+kp$ for any integers k, k'.) Therefore G_p has a conjugate X such that $G'_p \leqslant N(X)$. By Corollaries 7.12 and 7.19, $X = G'_p$, and it follows that G'_p is conjugate to G_p.

Parts (ii) and (i) of the theorem follow in succession. □

COROLLARY 7.24 (Sylow's second theorem). *In a finite group the Sylow p-subgroups (for a fixed prime p) are all conjugate and therefore isomorphic.*

Proof. Sylow p-subgroups exist. Clearly they are finite in number. Therefore Theorem 7.23 (ii) implies that they are all conjugate. □

COROLLARY 7.25. *Let G be a finite group and let G_p be any Sylow p-subgroup of G. If P is any p-subgroup of G then P lies in a suitable conjugate of G_p.*

Proof. It is clear that P lies in some Sylow p-subgroup G'_p of G. By the previous corollary $G'_p = G_p^x$ for a suitable element x of G; so $P \leqslant G_p^x$, as desired. □

COROLLARY 7.26. *Every p-subgroup of a finite group G lies in some subgroup of order p^α, if $|G| = p^\alpha r$ and p is prime to r.* □

COROLLARY 7.27 (Sylow's third theorem). *In a finite p-group the number of Sylow p-subgroups (for a fixed prime p) is congruent to 1 modulo p.*

Proof. This follows from Theorem 7.23 (i). □

We observe that Corollaries 7.16–7.21 may now be restated for finite groups and for $\pi = \{p\}$, with the aid of the Sylow theorems. We leave this task to the reader, but we shall write down the appropriate forms of Theorem 7.13 and Corollary 7.14.

COROLLARY 7.28. *If the group G has a finite normal subgroup K and if K_p is a Sylow p-subgroup of K, then $G = KN(K_p)$.* □

COROLLARY 7.29. *If the group G has a finite subgroup K containing the normalizer of a Sylow p-subgroup of G, then $K = N(K)$.* □

The next result is important for Chapter 9.

COROLLARY 7.30. *If G is a finite group in which every maximal subgroup is normal, then any Sylow p-subgroup (p being an arbitrary prime) of G is normal in G.*

Proof. A maximal subgroup of a group G is defined to be a subgroup S such that $S < G$, and $S \leqslant T \leqslant G$ implies $S = T$ or $T = G$. It is clear that a finite group always has at least one maximal subgroup or is trivial.

Suppose that the group G, which satisfies the hypothesis of the corollary, has a Sylow p-subgroup G_p for which $N(G_p) < G$. There is then a maximal subgroup M of G, containing $N(G_p)$. Corollary 7.29 indicates that $M = N(M)$, contradicting the fact that M is normal in G. Therefore all Sylow p-subgroups of G are normal. □

Theorem 7.23 imposes the strong restriction (in the context of infinite groups) that the Sylow p-subgroups are finite in number. It is possible to construct groups with an infinite number of Sylow p-subgroups, non-isomorphic ones among them. Instead of doing this we content ourselves with what is, in the circumstances, a rather mild example.

Example 7.31. Let $G = \prod_{n>0}^{D} G_n$ where G_n is isomorphic to the symmetric group S_3 for each n. Let a_n, b_n be the elements of G_n which correspond to (12), (13) respectively in S_3. Put

$$A = \prod_{n>0}^{D} \mathrm{gp}\{a_n\}, \qquad B = \prod_{n>0}^{D} \mathrm{gp}\{b_n\}.$$

An easy verification shows that if $g \notin A$ then $\mathrm{gp}\{A, g\}$ contains elements of order 3; this means that A is a Sylow 2-subgroup of G. So is B, for similar reasons. Now A and B are certainly isomorphic, but they are not conjugate in G. For if $A^x = B$ for some x in G (that is, x is a function from $\{n\}$ to $\{G_n\}$), then $nx \neq 1$ for every n since $a_n \neq b_n$; it follows that x could not have finite support, and therefore does not exist.

The final topic of this chapter is of a slightly different type from what has gone before. We shall prove an existence theorem in the case of a finite group having a *normal* subgroup with coprime order and index; the assumption of normality is made to appear reasonable by the situation in A_5, which is, of course, a simple group. The proof falls into two cases, the first of which is essentially due to Schur.

THEOREM 7.32. *Let G be a group with an abelian normal subgroup K such that G/K has order n and $x^m = 1$ for each element x of K, m and n being coprime positive integers. Then G has a subgroup C such that $G = KC$ and $K \cap C = 1$.*

Proof. We shall denote elements of G and of its abelian subgroup K by small Roman letters, as usual, and elements of the finite group G/K by small Greek letters. In each element α of G/K we choose an element g_α; thus g_α is a fixed element (depending on α) of G.

We see that $g_\alpha g_\beta$ and $g_{\alpha\beta}$ lie in the same coset of K in G. Therefore

$$g_\alpha g_\beta = g_{\alpha\beta} h_{\alpha,\beta}, \tag{1}$$

where $h_{\alpha,\beta}$ is some element of K. We now derive a relation between such elements by using the associative law in G. We obtain from (1):

$$g_\alpha(g_\beta g_\gamma) = g_\alpha g_{\beta\gamma} h_{\beta,\gamma} = g_{\alpha\beta\gamma} h_{\alpha,\beta\gamma} h_{\beta,\gamma},$$

$$(g_\alpha g_\beta) g_\gamma = g_{\alpha\beta} h_{\alpha,\beta} g_\gamma = g_{\alpha\beta} g_\gamma h^\gamma_{\alpha,\beta} = g_{\alpha\beta\gamma} h_{\alpha\beta,\gamma} h^\gamma_{\alpha,\beta}.$$

By $h^\gamma_{\alpha,\beta}$ we mean $g_\gamma^{-1} h_{\alpha,\beta} g_\gamma$. It follows that

$$h_{\alpha,\beta\gamma} h_{\beta,\gamma} = h_{\alpha\beta,\gamma} h^\gamma_{\alpha,\beta}. \tag{2}$$

We now take the product on each side of (2) as α varies over all the elements of the finite group G/K. Notice that the order of the factors is irrelevant because K is abelian. If we define c_β by

(3) $$c_\beta = \prod_{\alpha \in G/K} h_{\alpha,\beta}$$

then we obtain from (2)

(4) $$c_{\beta\gamma} h^n_{\beta,\gamma} = c_\gamma c^\gamma_\beta$$

where c^γ_β means $g_\gamma^{-1} c_\beta g_\gamma$. It should be observed that as α runs through the elements of G/K so does $\alpha\beta$, provided β is fixed.

Since m and n are coprime, we can find an integer n' such that

$$nn' \equiv 1 \text{ modulo } m.$$

If we put $d_\alpha = c_\alpha^{-n'}$ then (4) gives

(5) $$d^\gamma_\beta d_\gamma = d_{\beta\gamma} h^{-1}_{\beta,\gamma}$$

for each pair β, γ in G/K. Note that the abelian property of K is used again, and that the order of $h_{\beta,\gamma}$ divides m.

Next we define a set C of elements of G:

$$C = \{g_\alpha d_\alpha : \alpha \in G/K\}.$$

We have

$$\begin{aligned}(g_\alpha d_\alpha)(g_\beta d_\beta) &= g_\alpha g_\beta d^\beta_\alpha d_\beta \\ &= g_{\alpha\beta} h_{\alpha,\beta} d^\beta_\alpha d_\beta \\ &= g_{\alpha\beta} d_{\alpha\beta}\end{aligned}$$

by (1) and (5). It follows that C is isomorphic to G/K and also that C is a subgroup of G. The required facts that $G = KC$ and $K \cap C = 1$ are now clear. □

COROLLARY 7.33. *If K in Theorem 7.32 is central in G then $G \cong K \times C$.*

Proof. Since K and C commute element by element, the direct product criterion of Theorem 4.39 gives this result with very little difficulty. □

We pass on to the case in which K need not be abelian. The proof here is due to Zassenhaus, whence the name 'Schur–Zassenhaus theorem' for the final result.

THEOREM 7.34. *Let G be a group with a normal subgroup K such that G/K and K have coprime orders n and m respectively. Then G has a subgroup of order n.*

Proof. We proceed by induction on $|G|$. We take $m > 1$ since there is nothing to prove if $m = 1$. Theorem 7.32 then starts off the induction.

It may be advantageous to illustrate the general step by a diagram showing relations between subgroups of G. We shall not explain this in detail. It is, perhaps, enough to remark that the lines near K and N indicate that $G = \text{gp}\{K, N\}$ and that $K \cap N$ is the largest subgroup common to K and N.

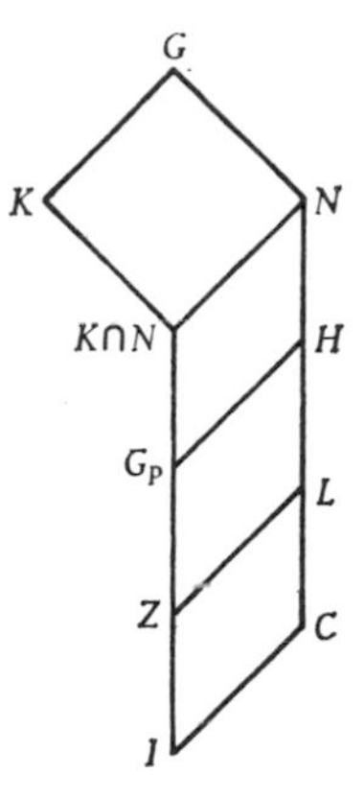

Let p be any prime divisor of m. Since m and n are coprime, K contains a Sylow p-subgroup G_p of G according to Sylow's first theorem. Let N denote the normalizer of G_p in G. Since $K \trianglelefteq G$, we see that K contains all the conjugates of G_p in G according to Sylow's second theorem. Now $N \cap K$ is the normalizer of G_p in K. It follows from Corollary 4.14 that

$$|G:N| = |K:N \cap K|.$$

It is an elementary fact about indices that

$$|G:N||N:N \cap K| = |G:K||K:N \cap K|.$$

These two displayed equations give at once that

$$|N:N \cap K| = |G:K|.$$

Because $K \trianglelefteq G$, Corollary 6.02 shows that $N \cap K \trianglelefteq N$. We know that $G_p \trianglelefteq N$. It follows from the third isomorphism theorem that

$$N/(N \cap K) \cong (N/G_p)/((N \cap K)/G_p).$$

Now $|N/G_p| < |G|$, while $|N:N \cap K| = n$ as above. The induction hypothesis may be applied to N/G_p, and we conclude that G has a subgroup H such that H/G_p is of order n.

If Z is the centre of G_p then $Z > 1$, by Theorem 4.29. Now Z is a characteristic subgroup of G_p by Theorem 4.34, and $G_p \trianglelefteq N$; so Z is normal in N by Theorem 4.35. We have $Z \trianglelefteq H$ because $Z \leqslant H \leqslant N$. The third isomorphism theorem gives

$$H/G_p \cong (H/Z)/(G_p/Z).$$

Since H/G_p has order n while G_p/Z has order a proper factor of $|G_p|$, the induction hypothesis applies again. Thus H/Z has a subgroup, L/Z say, of order n.

Schur's Theorem 7.32 shows that L has a subgroup C of order n. This is a subgroup of G with the desired property. □

Problems

1. Prove that there is no simple group of order 700.

2. How many subgroups of orders 2, 3, 4, 5, 6, 8, respectively, does the symmetric group S_4 contain?

3. Find all the Sylow 2-subgroups of the infinite dihedral group (described in Problem 5 of Chapter 3). Show that they fall into two conjugacy classes.

4. The group G has order pq, where p and q are distinct primes. Prove that G has a proper normal subgroup. Prove further that if neither prime is congruent to 1 modulo the other then G is abelian.

5. Prove that no pair of distinct Sylow π-subgroups of the subgroup H of the group G can lie in the same Sylow π-subgroup of G.

6. The finite group G has an automorphism α of prime order p. Prove that α must leave fixed at least one Sylow p-subgroup of G.

7. Describe a homomorphism from any group G into the symmetric group on the right cosets of its subgroup H. Prove that if H has finite index in G then the kernel of the homomorphism has finite index in G. Deduce the following:

   (i) If G is a p-group and H is maximal in G then H is normal in G.
   (ii) If H has index p in G, and G is otherwise arbitrary, then there is a right transversal of H in G with the form $\{g^i : 0 \leqslant i < p\}$.

*8. Prove that the right regular representation of a finite group G contains an odd permutation if and only if a Sylow 2-subgroup of G is cyclic and non-trivial.

Deduce that a group G with the condition mentioned cannot be non-abelian simple.

Prove further that if n is the largest odd factor of $|G|$ then G has a subgroup of order n.

9. The group G has order p^2q where p, q are distinct primes, and P_1, P_2 are two distinct Sylow p-subgroups in it. Prove that $P_1 \cap P_2$ is a normal subgroup (possibly trivial) of G.

Deduce that G cannot be a simple group.

10. The group $\mathscr{M}$ is generated by

$$A = \begin{pmatrix} 0 & 1 \\ 1 & -1 \end{pmatrix}, \qquad B = \begin{pmatrix} -1 & 1 \\ 1 & 0 \end{pmatrix},$$

the entries of the matrices being residue classes modulo 3. Calculate A^3 and $B^{-2}AB^2$. Prove that $\mathrm{gp}\{A, B^2\}$ has order 16 and that $\mathrm{gp}\{AB\}$ has order 3. Hence, or otherwise, find the order of $\mathscr{M}$.

11. The group G has a finite number of (non-trivial) Sylow p-subgroups for some prime p. Let $\mathscr{C}$ be the set of them. Show that there is a homomorphism from G into the symmetric group on $\mathscr{C}$. Hence prove the following statements:

(i) G has a normal p-subgroup P which has finite index in every member of $\mathscr{C}$.

(ii) If G has precisely $1+p$ Sylow p-subgroups then P has index p in every member of $\mathscr{C}$.

(iii) A finite group with precisely $1+p$ Sylow p-subgroups is non-simple or has a Sylow p-subgroup of order p.

12. Use the result of the preceding problem to prove that there is no simple group of order less than 60.

13. Let G be a finite group with a subgroup S of index n, where $n > 1$, and define a homomorphism of G into the symmetric group on the right cosets of S in G. Hence prove the following statements.

(i) If G is simple then the order of G divides $n!$

(ii) If $n = 2$ or 3 or 4 then G cannot be (non-abelian) simple.

(iii) There is a finite (non-abelian) simple group with a subgroup of index 5.

(iv) If p is the smallest prime factor of the order of G then any subgroup of index p in G is normal in G.

14. The group G has a normal subgroup N, and H is minimal in the set of those subgroups of G for which $G = NH$. Prove that every Sylow p-subgroup of $N \cap H$ is normal in H.

****15.** The group G has order 60. Find the possible numbers of its Sylow p-subgroups (for $p = 2, 3, 5$) permitted by Sylow's third theorem.

Prove that if G is simple, then there cannot be 3 Sylow 2-subgroups or 4 Sylow 3-subgroups; and that if G is not isomorphic to A_5 then there cannot be 5 Sylow 2-subgroups.

Deduce that a simple group (other than A_5) of order 60 would have precisely 15 elements of order 2 and 15 Sylow 2-subgroups. Derive a contradiction by considering the centralizer of an element lying in more than one Sylow 2-subgroup, and conclude that A_5 is the only simple group of order 60.

*16. Prove the true statements among the following and disprove the false.

(i) If P is a Sylow p-subgroup of the finite group G then $P \cap H$ is a Sylow p-subgroup of the subgroup H of G.

(ii) If P is a Sylow p-subgroup of the finite group G then $P/(P \cap N)$ is isomorphic to a Sylow p-subgroup of the factor group G/N of G.

(iii) There is a finite abelian group which has non-isomorphic Sylow π-subgroups, for a set of primes π.

(iv) Any Sylow p-subgroup of $G_1 \times G_2$ is the direct product of Sylow p-subgroups of G_1 and G_2.

(v) If the group G has order $p^\alpha r$ (where p is prime, $\alpha \geqslant 1$, and r is not divisible by p) then all subgroups of order $p^{\alpha-1}$ are isomorphic.

*17. Let the finite group G of order n have a subgroup H of index r. Show that there are at most $n-r+1$ elements in the union of all the conjugates of H. Deduce that if each element of G lies in some conjugate of H then $H = G$.

Suppose that the prime p divides $|G|$ but does not divide $|G:C(g)|$ for any $g \in G$. Prove that a Sylow p-subgroup of G is central in G.

**18. The finite group G has an abelian normal subgroup K with index n and order m, these being coprime integers. Let C be a subgroup of G with order n. Prove that if $G = KD$ for some subgroup D of G then a suitable conjugate of C lies in D. (Hint: first consider the case in which $K \cap D = 1$.)

Appendix to Chapter 7

IN recent years interesting new proofs of Sylow's three classical theorems have been found. They depend slightly on the theory of permutation groups and are otherwise elementary, but they seem no easier to the novice than the classical proofs. We give the details in this appendix, but they will not be used in the rest of the book.

Our total formal requirement of the theory of permutation groups consists of two definitions and a consequential theorem. Let a set Δ be given, and a group G of permutations of Δ, and a subset Γ of Δ.

DEFINITION. The *stabilizer* of Γ in the situation just described is the subset $\{g : \Gamma g = \Gamma\}$ of G; that is, the set of all elements of G which leave Γ fixed *as a set* (not necessarily element by element),

Now the stabilizer S of Γ is a subgroup, for the following reasons. Clearly S contains 1. And if $x, y \in S$, so that

$$\Gamma x = \Gamma, \qquad \Gamma y = \Gamma,$$

then $\Gamma x y^{-1} = \Gamma$, and $xy^{-1} \in S$. The usual subgroup criterion (Theorem 3.06) proves the assertion about S.

Next suppose that Γ is a one-element set, $\{\gamma\}$ say, so that S is $\{g : \gamma g = \gamma\}$. Let x, y be any two elements of G for which $\gamma x = \gamma y$. Then $xy^{-1} \in S$. By the very definition of coset, this means that x and y lie in the same right coset of S in G. Conversely, suppose that x and y are so situated in G; then $x = sy$ for some $s \in S$, and $\gamma x = \gamma s y = \gamma y$. Therefore the possible images in Δ of γ, under the action of G, are in one–one correspondence with the right cosets of S in G; and so the number of these images, being the index of S in G, divides the order of G when the latter is finite, by the theorem of Lagrange.

DEFINITION. The *transitivity class* (or *orbit*) of γ in Δ under G, in the above context, is the set of images $\{\gamma g : g \in G\}$ of γ.

We have now proved our elementary theorem about permutation groups.

THEOREM. *If G is a finite permutation group on a set Δ then the number of elements in each transitivity class under G divides the order of G.* □

We are ready to prove facts about Sylow structure.

THEOREM. *Let G be a group of order $p^{\alpha} r$ where $\alpha > 0$, p is prime, and r is not divisible by p. Then G contains at least one subgroup of each order p^{β} where $0 \leqslant \beta \leqslant \alpha$.*

Proof. We consider all complexes of G with p^β elements, with the intention of proving one of them to be a subgroup. Let $\mathscr{C}$ be the set of all such complexes. The number of elements of $\mathscr{C}$ is of course $\binom{p^\alpha r}{p^\beta}$, the number of combinations of $p^\alpha r$ objects taken p^β at a time.

We define a group of permutations on $\mathscr{C}$ as follows. For each $g \in G$, let $g\phi$ be the permutation of $\mathscr{C}$ given by $C . g\phi = Cg$ for all C in $\mathscr{C}$; here Cg is, of course, the complex product. To check that $g\phi$ *is* a permutation is easy; $g\phi$ is onto $\mathscr{C}$ because $(Cg^{-1}).g\phi = C$, and $g\phi$ is one–one because $Cg = C'g$ implies $C = C'$. We have

$$(g_1\phi)(g_2\phi) = (g_1 g_2)\phi$$

for all g_1, g_2 in G since

$$(C.g_1\phi)(g_2\phi) = (Cg_1)(g_2\phi) = (Cg_1)g_2 = C(g_1 g_2) = C.(g_1 g_2)\phi.$$

Closure of $\{g\phi : g \in G\}$ is immediate, and the associative law is easy; 1ϕ acts as an identity element, and $g^{-1}\phi$ is an inverse of $g\phi$. Hence $\{g\phi : g \in G\}$ is a group, H say. Moreover, we have shown that ϕ is a homomorphism from G onto H.

Take an arbitrary element x of G and an arbitrary transitivity class $\mathscr{T}$ of $\mathscr{C}$. We assert that x lies in a suitable element of $\mathscr{T}$. Take $C \in \mathscr{T}$, and take $c \in C$; then $1 = cc^{-1} \in Cc^{-1}$, and $x \in Cc^{-1}x$, and of course $Cc^{-1}x \in \mathscr{T}$, as required.

This transitivity class $\mathscr{T}$ need not, of course, form a partition of G since one element of G may lie in several elements of $\mathscr{T}$. But we can say that $\mathscr{T}$ contains at least $p^{\alpha-\beta}r$ elements of $\mathscr{C}$, because each C in $\mathscr{C}$ has only p^β elements while G itself has $p^\alpha r$ elements. Further, if $\mathscr{T}$ contains precisely $p^{\alpha-\beta}r$ elements of $\mathscr{C}$ then the elements of $\mathscr{T}$ are all disjoint (as complexes of G) and we do have a partition of G.

In fact, we can prove more under the assumption that $\mathscr{T}$ contains just $p^{\alpha-\beta}r$ elements. Some element of $\mathscr{T}$, say the complex C, contains 1 (in accordance with a fact proved above). Since $\mathscr{T}$ is a partition of G only one complex (namely C) of $\mathscr{T}$ contains 1. Suppose $x \in C$. Then $1 \in Cx^{-1}$, which also lies in the transitivity class $\mathscr{T}$. It follows that $C = Cx^{-1}$. If y is another arbitrary element of C, we deduce from this that $yx^{-1} \in C$. But now our subgroup criterion applies: C is in fact a subgroup. And clearly $\mathscr{T}$ consists of the subgroup C and its right cosets in G. Conversely, the set of right cosets of a subgroup K in G forms a transitivity class $\mathscr{T}$ of $\mathscr{C}$ with $p^{\alpha-\beta}r$ elements, if K has order p^β.

The topic of immediate interest, then, is the number t of transitivity classes $\mathscr{T}$ of $\mathscr{C}$ containing precisely $p^{\alpha-\beta}r$ elements of $\mathscr{C}$; the assertion of the theorem follows from the fact $t > 0$, which we now try to prove. Consider transitivity classes $\mathscr{U}$ of $\mathscr{C}$ containing more than $p^{\alpha-\beta}r$ elements of $\mathscr{C}$. How many elements does such a class $\mathscr{U}$ contain? This number must divide the order of H, by the previous theorem, and so the order of G, because there is a homomorphism from G onto H; and, being a factor of $p^{\alpha}r$ which exceeds $p^{\alpha-\beta}r$, it must be divisible by $p^{\alpha-\beta+1}$. This fact suffices for our purpose—it shows that the total number of elements of $\mathscr{C}$ each lying in a transitivity class $\mathscr{U}$ is $kp^{\alpha-\beta+1}$ for some integer k.

We now enumerate the elements of $\mathscr{C}$ according to whether they lie in a class like $\mathscr{T}$ or a class like $\mathscr{U}$. (Of course, no transitivity class contains *less* than $p^{\alpha-\beta}r$ elements of $\mathscr{C}$.) We obtain

$$\binom{p^{\alpha}r}{p^{\beta}} = tp^{\alpha-\beta}r+kp^{\alpha-\beta+1};$$

that is,

$$\binom{p^{\alpha}r}{p^{\beta}} \equiv tp^{\alpha-\beta}r \quad (\text{modulo } p^{\alpha-\beta+1}).$$

The essential point in the present proof is to observe that this congruence is largely independent of the structure of G. In fact, all we have used is that G has order $p^{\alpha}r$ and has t transitivity classes $\mathscr{T}$ with just $p^{\alpha-\beta}r$ elements of $\mathscr{C}$. Therefore the congruence will hold for *any* group G of order $p^{\alpha}r$, provided that we regard t as a function of G.

We may, therefore, choose any group G_0 of order $p^{\alpha}r$ for G; we take G_0 to be the cyclic group of that order. This group exists for every order $p^{\alpha}r$ and has one and only one subgroup of order p^{β}, and so one and only one transitivity class $\mathscr{T}$ with $p^{\alpha-\beta}r$ elements; $t = 1$. The assertion just made for cyclic groups follows from the theory of abelian groups developed in Chapter 5 (for instance). Substitution of $G = G_0$ and $t = 1$ therefore gives

$$\binom{p^{\alpha}r}{p^{\beta}} \equiv p^{\alpha-\beta}r \quad (\text{modulo } p^{\alpha-\beta+1}).$$

It follows that, for general G and corresponding t,

$$tp^{\alpha-\beta}r \equiv p^{\alpha-\beta}r \quad (\text{modulo } p^{\alpha-\beta+1}).$$

Since p does not divide r, which may therefore be cancelled, this congruence shows that $t \neq 0$. Thus $t > 0$, and the theorem is proved. □

COROLLARY (Sylow's first theorem). *If the group G has order $p^{\alpha}r$, where the prime p does not divide r, then G has at least one subgroup of order p^{α}.*

THEOREM. *Let G be a group of order $p^{\alpha}r$ where $\alpha > 0$, p is prime, and r is not divisible by p. Then the number of subgroups in G of each order p^{β}, where $0 \leqslant \beta \leqslant \alpha$, is congruent to 1 modulo p.*

Proof. The analysis which gave the first Sylow theorem also gives this theorem with little further development. In the course of the earlier proof we showed that G has t subgroups of order p^{β}, where t satisfies the congruence

$$tp^{\alpha-\beta} \equiv p^{\alpha-\beta} \quad (\text{modulo } p^{\alpha-\beta+1}).$$

This, of course, implies that $t \equiv 1$ modulo p. The theorem follows. □

COROLLARY (Sylow's third theorem). *The number of Sylow p-subgroups of the finite group G is congruent to* 1 *modulo p.* □

THEOREM. *If G_p is a Sylow p-subgroup of the finite group G, and if P is any p-subgroup of G, then G_p contains a suitable conjugate of P.*

Proof. We suppose as usual that G_p has order p^α and that the set $\mathscr{R}$ of its right cosets in G has r elements. We define a certain group of permutations which act on $\mathscr{R}$, as follows. If $x \in P$ then let $x\phi$ map $G_p g$ in $\mathscr{R}$ onto $G_p gx$ in $\mathscr{R}$. We check that $x\phi$ is a permutation of $\mathscr{R}$. The equation

$$(G_p x^{-1})(x\phi) = G_p x^{-1}x = G_p$$

ensures that $x\phi$ is onto $\mathscr{R}$. Since $G_p gx = G_p hx$ implies $G_p g = G_p h$, we see that $x\phi$ is one–one. Therefore $x\phi$ is a permutation.

Next, we assert that $(x\phi)(y\phi) = (xy)\phi$ for all x, y in P because it is easy to verify that each of $(x\phi)(y\phi)$ and $(xy)\phi$ maps the coset $G_p g$ onto the coset $G_p gxy$. It follows readily that $\{x\phi : x \in P\}$ is a group K of permutations of $\mathscr{R}$. Its order clearly divides $|P|$ and is therefore a power of p, say p^β.

Consider now the transitivity classes in $\mathscr{R}$ under K. The first theorem of this Appendix shows that each transitivity class contains p^γ cosets, for some γ satisfying $0 \leqslant \gamma \leqslant \beta$. But the number r of elements in $\mathscr{R}$ is not divisible by p, and so we must have some $\gamma = 0$. In other words, there is an element $G_p g$ of $\mathscr{R}$ which is fixed under the action of P; we have $G_p gx = G_p g$ for all x in P. Since $G_p gxg^{-1} = G_p$ we see that $gxg^{-1} \in G_p$ for all $x \in P$; that is, $gPg^{-1} \leqslant G_p$. That is precisely what had to be proved. □

COROLLARY (Sylow's second theorem). *Any two Sylow p-subgroups of a finite group are conjugate.* □

8

Generators and relations

THE free groups are an important and in many ways a natural class of groups to consider. They arise in problems from branches of mathematics other than algebra, and they are closely related to several aspects of group theory.

Our approach to free groups will be through generators and relations. Any group contains a set of generators (as we saw in Chapter 3), for the set of all its elements will always generate the group. There will, of course, be many relations involving these generators; for instance, relations of the forms $ab = 1$, $a^n = 1$. The quaternion group Q_8 gives another illustration (see Example 1.30). If we take a, b to be the matrices

$$\begin{pmatrix} 0 & 1 \\ -1 & 0 \end{pmatrix}, \qquad \begin{pmatrix} 0 & i \\ i & 0 \end{pmatrix}$$

respectively, it will be found that $Q_8 \cong \mathrm{gp}\{a, b\}$ and that

$$a^2 = b^2 = (ab)^2$$

is a pair of relations involving these generators. Finally consider the group $D = \mathrm{gp}\{a, b\}$ of Chapter 3, problem 5. Here

$$a = \begin{pmatrix} 1 & 0 \\ 0 & -1 \end{pmatrix}, \qquad b = \begin{pmatrix} 1 & 1 \\ 0 & -1 \end{pmatrix},$$

and we have $a^2 = b^2 = 1$ as relations involving the generators.

Some experimenting with Q_8 and D may perhaps make the reader feel that all relations in the given generators can be derived from the stated relations and no others, and convince him that it is natural to think of Q_8 and D as being somehow 'defined' by the above relations. We thus have a problem converse to that of finding relations among generators of a *given* group; this converse asks if a given set of relations can be used to define a group in some abstract fashion. This is the sort of problem which arises, for example, in topology and which we shall answer once we have put the question into a rigorous framework—the theory of free groups.

Another example may help the reader to see the difficulties we face. Suppose we feel a need to construct a group with generators a, b and with the relation $a^2 = b^2$. In what sense does such a group exist? There certainly are such groups—the trivial group will satisfy any relation, and this particular relation is also satisfied in both Q_8 and D. What can be said about *all* groups associated with this relation? And is there a special, distinguished group in this set?

We shall first define groups 'with no relations'—the free groups—and then show how to introduce such relations as we desire into these groups. There must, of course, be equations like $xx^{-1} = 1$ in any group, but it is best to regard these as not being proper relations.

We take any non-empty set X and proceed to define the free group on X. Let $X = \{x_\lambda : \lambda \in \Lambda)$ where Λ is a suitable index set. We take another set in one–one correspondence with X; call it X^{-1} and write its elements as x_λ^{-1}, for $\lambda \in \Lambda$. Note that the superscript has nothing to do with inversion in a group, as we have no group in which to operate yet—though misleading at the moment this particular symbolism will be convenient later.

DEFINITION. A *word* in the elements of $X \cup X^{-1}$ is an ordered set of n elements, each from $X \cup X^{-1}$ with repetitions allowed, for some $n \geqslant 0$. The *length* of a word is the integer n.

Notation for words will also be suggestive in a formal way. A typical word of length $n > 0$ will be written as $x_{\lambda_1}^{\epsilon_1} \dots x_{\lambda_n}^{\epsilon_n}$ where each ϵ_i is 1 or -1; the (unique) word of length 0 will be written as 1. (Of course, x^1 means x.) Note carefully that multiplication of elements in $X \cup X^{-1}$ is not implied by the notation. A case could be made at this stage for writing the above word as $(x_{\lambda_1}^{\epsilon_1}, x_{\lambda_2}^{\epsilon_2}, \ldots, x_{\lambda_n}^{\epsilon_n})$, for instance. As it is an ordered set it could be regarded as a sort of vector.

Example 8.01. If $X = \{x, y\}$ then some typical words in $X \cup X^{-1}$ are:

$$x^{-1}, \quad x^{-1}xyy^{-1}, \quad x^{-1}xx^{-1}, \quad yyyxy^{-1}, \quad 1.$$

It is a consequence of the definition that two words are equal if and only if they have the same length and corresponding terms are equal. The words above are a set of five distinct words, for no two have the same length.

Next we define the 'product' of two words. First, let w be an arbitrary word; then the product of 1 with w and of w with 1 is to be w. Secondly, take two words of positive lengths, say $u = x_{\lambda_1}^{\epsilon_1} \ldots x_{\lambda_n}^{\epsilon_n}$ and $v = x_{\mu_1}^{\delta_1} \ldots x_{\mu_m}^{\delta_m}$ in the obvious notation. The product uv is defined to be

$$x_{\lambda_1}^{\epsilon_1} \ldots x_{\lambda_n}^{\epsilon_n} x_{\mu_1}^{\delta_1} \ldots x_{\mu_m}^{\delta_m}.$$

Note that the length of the product is $n+m$, and that $uv \neq vu$ in general. The elements of $X \cup X^{-1}$ are words of length 1, and every word (except perhaps the empty word) is the product of certain of these particular words. Of course, the set of all words will not form a group under this multiplication, for no word of positive length will have an inverse. But it is clearly true that the associative law holds and that our set of words therefore forms a semigroup.

Our next aim is to define a group whose elements are certain equivalence classes of these words.

Definition. Two words u, v in $X \cup X^{-1}$ are *adjacent* if there are words z_1, z_2 and an element a in $X \cup X^{-1}$ for which $u = z_1 z_2$, $v = z_1 a a^{-1} z_2$; or $u = z_1 a a^{-1} z_2$, $v = z_1 z_2$. (Interpret $(x^{-1})^{-1}$ as x.)

Note that we cannot meaningfully talk of 'cancellation' because we are not working in a group. (In fact the very purpose of defining adjacent words is to enable us to talk of cancellation in due course.) In the examples above, $x^{-1}xyy^{-1}$ is adjacent to yy^{-1} (take $z_1 = 1$, $z_2 = yy^{-1}$, $a = x^{-1}$) and also to $x^{-1}x$ (take

$z_1 = x^{-1}x$, $z_2 = 1$, $a = y$), while $x^{-1}xx^{-1}$ is adjacent to x^{-1}. Of course, u is adjacent to v if and only if v is adjacent to u.

DEFINITION. Two words u, v in $X \cup X^{-1}$ are *equivalent* if there exist words $z_1,..., z_n$ (with $n \geqslant 1$) such that $u = z_1$, $v = z_n$, and z_i is adjacent to z_{i+1} for $i = 1,..., n-1$.

Two identical words are equivalent in a trivial way (with $n = 1$). In our examples $x^{-1}xyy^{-1}$ is equivalent to 1, for $x^{-1}xyy^{-1}$ is adjacent to yy^{-1} which is in turn adjacent to 1; and of course $x^{-1}xx^{-1}$ is equivalent to x^{-1}.

It is trivial to verify that our definition of equivalence between words is an equivalence relation (in the usual sense) on the set of all words in $X \cup X^{-1}$; the reflexive property is a triviality, symmetry follows from the definitions of equivalent and adjacent words, transitivity follows from the definition of equivalent words. We may therefore form the equivalence classes. That class containing the word w will be denoted by $[w]$.

DEFINITION. The *product* of equivalence classes $[u]$, $[v]$ of words in $X \cup X^{-1}$ is defined to be $[uv]$.

THEOREM 8.02. *The product of two equivalence classes of words in $X \cup X^{-1}$ is well defined, and the set of all equivalence classes with this binary operation forms a group.*

Proof. The statement about good definition is that $[u'v'] = [uv]$ if $[u'] = [u]$ and $[v'] = [v]$. The fact that u and u' are equivalent implies that $[u'v] = [uv]$, because $u'v$ and uv are adjacent if u' and u are adjacent. Similarly $[u'v'] = [u'v]$ follows from the equivalence of v and v'. Therefore $[u'v'] = [uv]$, as required.

We now verify the group axioms. Closure is clear from the definition of product and its subsequent justification. The associative law is a consequence of the associative law $u(vw) = (uv)w$ for the multiplication of the words u, v, w:

$$[u]([v][w]) = [u][vw] = [u(vw)],$$
$$([u][v])[w] = [uv][w] = [(uv)w],$$

and so
$$[u]([v][w]) = ([u][v])[w].$$

The class of words containing 1 is clearly an identity element, since $[w][1] = [w]$ for any word w. If $w = a_1 \ldots a_n$ with each $a_i \in X \cup X^{-1}$ then we may take $[a_n^{-1} \ldots a_1^{-1}]$ as an inverse of the class $[w]$, because $[wa_n^{-1} \ldots a_1^{-1}] = [1]$. □

Definition. The *free group* on the non-empty set X is the set of equivalence classes of words in $X \cup X^{-1}$ with the binary operation described above.

Definition. A word in $X \cup X^{-1}$ is *reduced* if it has the form $x_{\lambda_1}^{\epsilon_1} \ldots x_{\lambda_n}^{\epsilon_n}$ with $x_{\lambda_{i+1}}^{\epsilon_{i+1}} \neq x_{\lambda_i}^{-\epsilon_i}$ for $i = 1, 2, \ldots, n-1$. (Of course $x^{-(-1)}$ is to be understood as x.)

In Example 8.01 above x^{-1} and $yyyxy^{-1}$ and 1 are the reduced words.

Theorem 8.03. *Each equivalence class of words in $X \cup X^{-1}$ contains one and only one reduced word.*

Proof. Let w be a word in $X \cup X^{-1}$. There is no particular trouble in showing that if w is not reduced then it is adjacent to a word of smaller length, and so an inductive argument will produce an equivalent reduced word. To show that $[w]$ contains only one reduced word requires more thought, and is an essential step in the theory of free groups. We define a particular method of reduction, that is of obtaining a reduced word equivalent to a given word.

Let $w = a_1 \ldots a_n$ where $a_i \in X \cup X^{-1}$ for $1 \leqslant i \leqslant n$. Let $w_0 = 1$, $w_1 = a_1$. Suppose that w_i is defined, where $1 \leqslant i < n$; we produce the definition for w_{i+1}, in two separate cases. It must be remembered that w_i is a word, so that one can speak of its last term without ambiguity. If w_i does not have last term a_{i+1}^{-1} then we put $w_{i+1} = w_i a_{i+1}$; if w_i does end in a_{i+1}^{-1}, say $w_i = za_{i+1}^{-1}$, then z is uniquely determined (because $z_1 a_{i+1}^{-1} = z_2 a_{i+1}^{-1}$ implies that $z_1 = z_2$ by the definition of a word), and we define w_{i+1} to be z. This gives an inductive definition for $w_0, w_1, \ldots, w_n$; of course, if $n = 0$ then w_n is defined to be 1. A consequence of the definition is that w_i is reduced for $0 \leqslant i \leqslant n$, and that in particular w_n is reduced. Another easy fact is that w_i is equivalent to $a_1 \dots a_i$ for each i, and so $[w_n] = [w]$. Note also that if w is already reduced then $w_n = w$.

Next we show that two adjacent words u, v have reduced forms which are identical. Let

$$u = a_1 \dots a_r a_{r+1} \dots a_n,$$

$$v = a_1 \dots a_r x x^{-1} a_{r+1} \dots a_n$$

where $x \in X \cup X^{-1}$; this choice of u and v is general enough. Suppose the procedure described above gives the sequence $u_0 = 1, u_1, \dots, u_n$ for u, and $v_0 = 1, v_1, \dots, v_{n+2}$ for v. Since w_i was determined, in the inductive definition, by the first i factors in w we clearly have

$$u_0 = v_0, \quad u_1 = v_1, \quad \dots, \quad u_r = v_r.$$

We shall show that $u_r = v_{r+2}$, in two separate cases.

(i) If u_r does not end in x^{-1} then $v_r = u_r$, $v_{r+1} = v_r x$, and $v_{r+2} = v_r$. (Here we use the definition of w_i.) Therefore $u_r = v_{r+2}$.

(ii) If u_r does end in x^{-1}, say $u_r = zx^{-1}$, then z cannot have a reduced form $z_0 x$ for any z_0; for if it did then $u_r = z_0 x x^{-1}$ would not be in reduced form. Therefore $v_r = u_r$, $v_{r+1} = z$, and $v_{r+2} = zx^{-1} = u_r$, as required.

Thus in either case $u_r = v_{r+2}$. Since the final $n-r$ factors in the expressions for u and v correspond when taken in the obvious order, we have $u_{r+i} = v_{r+2+i}$ for $i = 0, 1, \dots, n-r$. In particular $u_n = v_{n+2}$, and it follows that the procedure for giving a reduced form yields the same result for u as for v.

Finally let u, v be any two reduced words in $[w]$. They can be associated by a chain of adjacent words, because they are equivalent. We apply our procedure to all the words in the chain. The result is the same reduced word in each case. Since the procedure does not alter words which are already reduced it will not alter u and v. Therefore u and v are identical words. Therefore $[w]$ contains precisely one reduced word. □

We shall not always write the elements of the free group on X as equivalence classes—we shall often merely use a representative from each class, usually the reduced word of the class, as no real confusion will result. In particular we write $[x]$ as x for each $x \in X \cup X^{-1}$, and $[1]$ as 1.

If the sets X and Y are in one–one correspondence then the free groups on them are (clearly) isomorphic. It is also true that if two free groups are isomorphic then the corresponding sets X and Y have the same cardinal (in other words, are in one–one correspondence); we shall prove this later for finite sets X and Y. Thus the cardinal of X is an invariant of the free group F on X, and it is called the *rank* of F.

We note some properties of free groups, which are easy enough to give us some feeling for these objects.

(i) Every free group is torsionfree. For if a is an element of finite order in the free group F on X then $a^n = 1$ for some $n \geqslant 1$. Let the reduced form of a be

$$b_r^{-1} \ldots b_1^{-1} a_1 \ldots a_s b_1 \ldots b_r$$

where $a_1 \neq a_s^{-1}$; of course, r may possibly be 0. Then

$$a^n = b_r^{-1} \ldots b_1^{-1}(a_1 \ldots a_s)^n b_1 \ldots b_r,$$

and so the reduced form of a^n has length $2r+ns$. But this length is 0 because $a^n = 1$. It follows that $r = s = 0$ since $n \geqslant 1$. Therefore the reduced form of a is the empty word, and $a = 1$.

(ii) In particular, if $X = \{x\}$ then the free group on X is infinite cyclic.

(iii) If X has more than one element then the free group on X is non-abelian, for if $x, y \in X$ then $xy \neq yx$.

It should be clear what the rigorous interpretation of the statement that a free group 'has no relations' is. It is simply the fact that if $a_1 \ldots a_n$ is a reduced word equal to 1, with each $a_i \in X \cup X^{-1}$, then $n = 0$ and the word is empty.

Next we consider a subtler concept, the free generators of a free group, and take as an illustration the free group F on the set $X = \{x, y\}$. The elements x, y, which certainly generate F, have the property that any reduced word in them which equals 1 must be the empty word. But other generators also have this property. Consider, for instance, $\{x, x^{-1}yx\}$. This set generates a subgroup of F which is F itself because it contains both x and y. A reduced word in x, y^x is the conjugate of a reduced word in x, y; for example, $x^{-1}(y^x)^2x^2(y^x)^{-1} = (x^{-1}y^2x^2y^{-1})^x$. Therefore the only reduced word in $\{x, y^x\}$ equal to 1 is the empty word.

DEFINITION. A *free basis* (or a *set of free generators*) for the free group F is a set of generators for F with the property that the only reduced word in them equal to 1 is the empty word. (Of course, a free basis is said to generate F freely.)

It should be clear that a free group has many bases other than the set on which it was defined. The concept is not applicable to arbitrary groups (for what would be meant by a free basis for the quaternion group?) In the above example $\{x, y^2\}$ freely generates a subgroup of F that happens (and can be proved) to be a proper subgroup of F; and $\{x, y, xy\}$ is a set of generators which is not a free basis. The length of a word depends strongly on the free basis in use; thus $x^{-1}yx$ has length 3 when referred to $\{x, y\}$ in the above example, and length 1 when referred to $\{x, x^{-1}yx\}$.

The next few results attempt to relate free groups to arbitrary groups.

THEOREM 8.04. *Let F be a free group with free basis*

$$X = \{x_\lambda : \lambda \in \Lambda\},$$

and let G be an arbitrary group. If $\{g_\lambda : \lambda \in \Lambda\}$ is a set of arbitrary elements of G, then there exists a unique homomorphism from F into G which maps x_λ onto g_λ, for all $\lambda \in \Lambda$.

Proof. Let ϕ be the mapping from X into G for which $x_\lambda \phi = g_\lambda$ $(\lambda \in \Lambda)$, and let $[w]$ be any element of F. We suppose that $w = x_{\lambda_1}^{\epsilon_1} \dots x_{\lambda_n}^{\epsilon_n}$, in the usual notation, and define $[w]\phi$ to be $g_{\lambda_1}^{\epsilon_1} \dots g_{\lambda_n}^{\epsilon_n}$. In order to show that ϕ is a mapping we have to show that if $[u] = [v]$ then $[u]\phi = [v]\phi$; but this is clear if u and v are adjacent, and it therefore holds when they are equivalent.

It is now easy to see that ϕ is a homomorphism. Let $[u]$, $[v]$ be arbitrary elements of F, so that $[uv] = [u][v]$, as above. Then the definition of ϕ should make it clear that $[uv]\phi = [u]\phi.[v]\phi$, and it follows that

$$([u][v])\phi = [u]\phi.[v]\phi$$

as required.

Since ϕ maps $[x_\lambda]$ onto g_λ, it is the required homomorphism. Its uniqueness follows from the fact that the images of the reduced words in $X \cup X^{-1}$ are determined once the images of the x_λ are specified. □

With the aid of this result we can discuss the invariance of rank.

THEOREM 8.05. *Let F_m, F_n be free groups of finite ranks m, n respectively. Then $F_m \cong F_n$ if and only if $m = n$.*

Proof. We mentioned earlier that if $m = n$ then $F_m \cong F_n$. Suppose, conversely, that $F_m \cong F_n$, and let G denote a group of order 2 with generator g. We consider homomorphisms from F_m to G, and for this purpose we suppose that F_m has free basis $\{x_1,\ldots, x_m\}$. Such a homomorphism is completely determined by the image of each x_i, and each image may be g or 1; there are therefore 2^m-1 distinct homomorphisms from F_m onto G, in addition to the trivial one in which every x_i is mapped onto 1.

The kernel K of a homomorphism from F_m onto G is a normal subgroup of F_m, and $F_m/K \cong G$ by the first isomorphism theorem. Conversely, a normal subgroup of index 2 in F_m is the kernel of a homomorphism of F_m onto an isomorphic copy of G. It follows that F_m has precisely 2^m-1 normal subgroups of index 2.

Similarly F_n has precisely 2^n-1 normal subgroups of index 2. Because $F_m \cong F_n$ we have $2^m-1 = 2^n-1$, and so $m = n$, as required. □

COROLLARY 8.06. *Every free basis for F_n contains just n elements.* □

Theorem 8.05 can be proved for free groups of arbitrary rank, but this requires an acquaintance with the necessary results concerning infinite sets.

Next, a converse to the theorem stating the existence of homomorphisms from a free group into an arbitrary group. (For the sake of tidiness we now *define* the free group on the empty set to be the group with one element.)

THEOREM 8.07. *If $G = \mathrm{gp}\{g_\lambda : \lambda \in \Lambda\}$ is a group with the property that, for an arbitrary group H containing a complex $\{h_\lambda : \lambda \in \Lambda\}$, the mapping $\theta : g_\lambda \to h_\lambda$ can be extended to a homomorphism of G into H, then G is a free group and $\{g_\lambda : \lambda \in \Lambda\}$ is a free basis for G.*

Proof. We make a particular choice for H; we take H to be the free group with $\{h_\lambda : \lambda \in \Lambda\}$ as a free basis. There is by Theorem 8.04 a homomorphism ϕ from H to G with $h_\lambda \phi = g_\lambda$ for all $\lambda \in \Lambda$. We are given, in addition, a homomorphism θ from G to H with $g_\lambda \theta = h_\lambda$. It is clear that

$$h_\lambda(\phi\theta) = h_\lambda, \qquad g_\lambda(\theta\phi) = g_\lambda,$$

and so $\phi\theta$, $\theta\phi$ are identity mappings on H, G respectively. It follows that θ is an isomorphism. Since H is free with free basis $\{h_\lambda : \lambda \in \Lambda\}$, it follows that G is free with free basis $\{g_\lambda : \lambda \in \Lambda\}$. □

Sometimes a free group is defined by the property mentioned in this theorem, but we have preferred a more constructive definition in the present account.

THEOREM 8.08. *Every group is isomorphic to a factor group of a suitable free group. Every group with a set of n generators is isomorphic to a factor group of the free group of rank n.*

Proof. Take a set of generators $\{g_\lambda : \lambda \in \Lambda\}$ for the arbitrary group G; if need be the set of all its elements may be taken. Let $X = \{x_\lambda : \lambda \in \Lambda\}$ be some abstract set in one–one correspondence with the generating set. Let X be a free basis for the free group F. Then there is (by Theorem 8.04) a homomorphism ϕ of F into G such that $x_\lambda \phi = g_\lambda$ for all $\lambda \in \Lambda$. The first statement of the theorem follows. For the second we simply take Λ to be a set of n elements. □

Example 8.09. Let F be the free group of (finite) rank n, and let A be the direct sum of n infinite cyclic groups. Then $F/K \cong A$, for some normal subgroup K of F, by Theorem 8.08. Since F/K is abelian, $\delta(F) \leqslant K$, by Theorem 4.21. Therefore, by the third isomorphism theorem, F/K is a factor group of $F/\delta(F)$ (compare Example 6.17). But $F/\delta(F)$ is an abelian group with a set of n generators. It is a consequence of the structure theorem (5.14) for finitely generated abelian groups that $F/\delta(F) \cong F/K$. We have therefore proved: if F is free of rank n then $F/\delta(F)$ is the direct sum of n infinite cyclic groups.

We mention in passing the concept of a *free abelian group*. This is simply the direct sum of some set of infinite cyclic groups, and we may define free basis and rank, if we wish, in the obvious way. The additive group of rationals is a torsionfree abelian group which is *not* free abelian, as the reader may prove for himself.

We are now ready for generators and relations. Suppose that we try to express the quaternion group as some factor group of a free group. We take a free group F with free basis $\{x, y\}$, and we take a quaternion group Q_8 generated by a, b where

$$a = \begin{pmatrix} 0 & 1 \\ -1 & 0 \end{pmatrix}, \qquad b = \begin{pmatrix} 0 & i \\ i & 0 \end{pmatrix}.$$

Everything is arranged once we define the homomorphism ϕ from F to Q_8 by putting

$$x\phi = a, \qquad y\phi = b.$$

Now we have the relations $a^2 = b^2 = (ab)^2$ in Q_8. They imply that y^2x^{-2} and $(xy)^2y^{-2}$ are in the kernel K of ϕ because

$$(y^2x^{-2})\phi = (y\phi)^2(x\phi)^{-2} = b^2a^{-2} = 1,$$
$$((xy)^2y^{-2})\phi = ((x\phi)(y\phi))^2(y\phi)^{-2} = (ab)^2b^{-2} = 1.$$

Thus the kernel contains the normal closure of $\{y^2x^{-2}, (xy)^2y^{-2}\}$—this is the least normal subgroup containing these two elements.

And now we have a hint as to what to do about defining relations for groups.

Let us take another point of view. Suppose that we want to *construct* a group with certain generators and relations, rather than investigate a given one. To be specific, let us not give ourselves a quaternion or any other group, but let us ask for a group with generators a, b and the relation $a^2 = b^2$. We take a free group F with free basis $\{x, y\}$. We take a normal subgroup K of F containing x^2y^{-2}, and we consider F/K. If $a = xK$, $b = yK$, then we certainly have $a^2 = b^2$ in F/K. If we take K to be the normal closure of x^2y^{-2} then we are justified in regarding the resulting group G/K as being (in some sense) the most comprehensive group with generators a, b and relation $a^2 = b^2$.

We now formalize these ideas.

Definition. A *group presentation* consists of a free basis X for a free group together with a set of words in $X \cup X^{-1}$.

We derive a group from a presentation in the following way. Let F be the free group on X, and let $W = \{w_\lambda : \lambda \in \Lambda'\}$ be the given set of words in $X \cup X^{-1}$, i.e. elements of F. Define K to be the normal closure in F of W. The group presented is then F/K. Since each $w_\lambda \in K$ we have, of course, $w_\lambda K = K$ in F/K, and so we may claim informally that 'the relation $w_\lambda = 1$ is satisfied in F/K'.

Thus in the above example $X = \{x, y\}$ and $W = \{x^2y^{-2}\}$.

A group presentation should ideally be denoted in some such way as $\mathrm{gp}\{X : W\}$. We shall, however, use a less exact notation in which 'relations' replace the words of W, for example $\mathrm{gp}\{X : w_\lambda = 1\ (\lambda \in \Lambda')\}$, as we shall usually be more interested in the group than in its presentation. The group discussed above would thus be written as $\mathrm{gp}\{a, b : a^2b^{-2} = 1\}$ or even as $\mathrm{gp}\{a, b : a^2 = b^2\}$.

Theorem 8.10 (von Dyck's theorem). *If the group G is defined by some set of relations and the group H is defined by a larger set of relations, then H is isomorphic to a factor group of G.*

Proof. It is understood that the generating sets of G, H are in one–one correspondence so that each is isomorphic to a suitable factor group of the same free group F. Thus $F/K_1 \cong G$, $F/K_2 \cong H$, where $K_1 \leqslant K_2$ by the condition on the relations. By the third isomorphism theorem,

$$H \cong F/K_2 \cong (F/K_1)/(K_2/K_1) \cong G/N$$

where $N \cong K_2/K_1$. Thus H is isomorphic to a factor group of G, as required. □

Example 8.11. Consider $G = \mathrm{gp}\{a, b : a^2 = b^2\}$. If we add the additional relation $b^2 = (ab)^2$ then we obtain the quaternion group Q_8, while if we add $b^2 = 1$ (and perhaps other relations, for all that we have proved so far) then we obtain the infinite dihedral group D. Thus von Dyck's theorem tells us that G has factor groups isomorphic to Q_8 and D; in particular, G is infinite.

Von Dyck's theorem completes the solution to the general problems about generators and relations posed earlier in this chapter. For we have seen that there is a distinguished group associated with given generators and relations, and every associated group is a factor group of the distinguished one.

In view of the distinction we have made between a presentation and the group defined by a presentation, it should be no surprise to find that one group may be presented in many ways. The example G above could be presented as

$$\mathrm{gp}\{a, b, c : a^2 = b^2,\ c = 1\}.$$

The less trivial instances which follow have been selected to show the sort of *ad hoc* calculation which is often the only effective way of investigating a given presentation.

Example 8.12. Consider $\mathrm{gp}\{a, b, c, d : bc = d,\ cd = a,\ da = b,\ ab = c\}$. If we eliminate d between the first and third relations in this group we find that $c = a^{-1}$. The relations then give $d = a^2$, $b = a^3$, and $b = a^{-2}$; thus $a^5 = 1$, and indeed $b^5 = c^5 = d^5 = 1$. It follows that the group has order 5 or less. But since there does exist a group of order 5 satisfying the above relations, namely the additive residues modulo 5 with $a = [1]$,

$b = [3]$, $c = [4]$, $d = [2]$, it follows that the group presented indeed has order 5. Note that this last step, establishing the *existence* of a particular group, is essential.

The point of this example is that the group presented may also be described as $\mathrm{gp}\{g : g^5 = 1\}$. The isomorphism is by no means evident from the presentations.

Example 8.13. Consider the following groups:

$$G = \mathrm{gp}\{a, b : a^4 = b^2,\ b^{-1}ab = a^{-1}\},$$
$$H = \mathrm{gp}\{c, d : cdcdc = dcd,\ dcdcd = cdc\}.$$

In G the element a^4 is central, being a power of both generators, and so

$$a^4 = b^{-1}a^4b = (b^{-1}ab)^4 = (a^{-1})^4,$$

from which we obtain $a^8 = 1$. Thus $\mathrm{gp}\{a\}$ is a normal subgroup of G, which has order at most 16. That G really does have order 16 must be proved by exhibiting a specific group satisfying the given relations. This may be achieved in several ways, for instance by means of the matrices

$$A = \begin{pmatrix} 3 & 3 \\ -3 & 3 \end{pmatrix}, \qquad B = \begin{pmatrix} 0 & 4 \\ 4 & 0 \end{pmatrix},$$

whose entries are residue classes modulo 17; the reader should be able to verify that

$$A^4 = B^2, \qquad B^{-1}AB = A^{-1},$$

and that $\mathrm{gp}\{A, B\}$ has order 16.

It is not obvious that $G \cong H$, but we shall proceed to prove this. In H we have

$$dcd = cdc.dc = dcdcd.dc,$$

and so $cddc = 1$, $c^2 = d^{-2}$. Therefore c^2 and d^2 are central elements in H. We then find that as $cdcdcd^{-1}c^{-1}d^{-1} = 1$ we have $(cd)^4 \in Z = \mathrm{gp}\{c^2\}$, and it may be seen that $\mathrm{gp}\{cdZ\}$ is a normal subgroup of H/Z. Hence H/Z has order at most 8. Another appeal to the defining relations gives

$$cdcd = c^2dcdc, \quad cdcd = dcdcd^2;$$

it follows that $c^2 = d^2$. But $c^2 = d^{-2}$, and we thus have $c^4 = 1$; Z has order at most 2. Therefore H has order 16 or less.

However, we know that there is a group G of order 16 with generators a, b and relations $a^4 = b^2$, $b^{-1}ab = a^{-1}$. If we substitute c for b and d for ab, we find that this group satisfies the given relations for H. Since $G = \mathrm{gp}\{b, ab\}$ we conclude that H has order 16, and that $G \simeq H$.

It is, of course, possible to consider presentations with infinitely many generators or relations, but as an arbitrary group has such a presentation there is some excuse to study a restricted class only.

DEFINITION. A *finite presentation of a group* is a group presentation consisting of a finite free basis and a finite set of words.

DEFINITION. A *finitely presented group* is a group which has a finite presentation.

All the examples considered explicitly above are finitely presented. Any finite group may be finitely presented because the set of its elements is finite and its multiplication table is finite.

DEFINITION. The *deficiency* of a finite presentation $\mathrm{gp}\{X : W\}$ of a group is the number of elements in X less the number of elements in W.

This definition of deficiency is good enough for our purpose, namely Theorem 8.16, but it may be enlightening to discuss how it could possibly be improved. Given any presentation $\mathrm{gp}\{X : W\}$, we may always decrease the deficiency by adding redundant elements to W; perhaps of the form w_1^x or $w_1 w_2$ where w_1, w_2 are already in W and $x \in X$. We might wish, conversely, to rid ourselves of such superfluous relations, but some tricky points arise.

Example 8.14. Consider $\text{gp}\{x : x^2 = 1,\ x^4 = 1,\ x^6 = 1\}$. This defines the group of order 2. We can change the presentation to $\text{gp}\{x : x^2 = 1\}$, and the deficiency becomes 0. But we could also delete one relation without changing the underlying group, obtaining $\text{gp}\{x : x^4 = 1,\ x^6 = 1\}$; in this case neither of the two remaining relations is dispensable, and the deficiency is -1.

One might compare this situation to that which arises when sets of generators for the cyclic group of order 6 are being examined. There is a one-element generating set, but there is also a two-element set of which each element generates a *proper* subgroup. We now consider some defining relations in these generators.

Example 8.15. Consider $\text{gp}\{x, y : x^2 = 1,\ y^3 = 1,\ [x, y] = 1\}$. This clearly defines the cyclic group of order 6. None of the three given relations is superfluous, for omission of any one gives rise to a larger group. If $x^2 = 1$ is omitted the resulting group has an infinite cyclic factor group; similarly if $y^3 = 1$ is omitted; while if $[x, y] = 1$ is abandoned we find a non-abelian factor group generated by the elements (12) and (123) in S_3.

Nevertheless this cyclic group of order 6 can be defined by *two* irredundant relations in the same irredundant generators: $\text{gp}\{x, y : x = (xy)^3,\ y = (xy)^4\}$. This is quite easy to prove.

A subtler way in which presentations of a group may be compared will be described. Suppose that the group G has a set of generators $\mathscr{G}$, and for the moment keep $\mathscr{G}$ fixed. We can then talk about 'relations in $\mathscr{G}$ which define G'; we mean a set W of words in X, with X in one–one correspondence with $\mathscr{G}$, and this correspondence inducing an isomorphism of $\text{gp}\{X : W\}$ with G. Suppose that $\mathscr{G}$ has m elements and that a smallest possible W has n elements. We might hope that we could prove (or that someone has proved) that $m - n$ is independent of $\mathscr{G}$ and therefore an invariant of G. This hope would be destroyed by examples which are too abstruse to describe here and whose existence helps to make comparison of presentations a frustrating task.

We now end our digression and regard deficiency as the concept specified in the formal definition, on page 163.

There are many examples of finite groups with negative deficiency, and some (such as the cyclic groups and the quaternion group) with zero deficiency; but none exist with positive deficiency.

THEOREM 8.16. *Any group defined by a finite presentation with positive deficiency is necessarily infinite.*

Proof. Let $G \cong F/K$ where F is free with free basis $\{x_1,..., x_m\}$ and K is the normal closure of the complex $\{w_1,..., w_n\}$ in F; we assume $m > n$. Define α_{ij} to be the sum of all the exponents of x_i appearing in w_j (taken in its reduced form if desired, though it makes no significant difference). Consider the n equations

$$\sum_{i=1}^{m} \alpha_{ij}\beta_i = 0 \quad (1 \leqslant j \leqslant n)$$

for the m numbers β_i. Because $m > n$ a well-known theorem† states that there exists a non-zero solution $\{\beta_1,..., \beta_m\}$. Though this solution is in the first instance a set of rational numbers (the α_{ij} being integers) we may clearly select it as a set of integers, not all zero.

We shall now construct a homomorphism from G into the infinite cyclic group C with generator c. First we define a mapping ϕ from $\{x_1,..., x_m\}$ into C by putting $x_i\phi = c^{\beta_i}$, for each i (note that β_i is an integer). Since F is free, ϕ defines a homomorphism from F into C, in accordance with Theorem 8.04. The choice of the α_{ij} and the β_i ensures that $w_j\phi = 1$ for $1 \leqslant j \leqslant n$. It follows that gp$\{w_1,..., w_n\}$ maps onto 1, and (more relevantly) that K maps onto 1 because K consists of products of *conjugates* of the $w_j^{\pm 1}$. Since $K\phi = 1$, there is a homomorphism ψ from F/K into C; it is defined by

$$(Kx)\psi = x\phi$$

† See, for instance, S. Perlis, *Theory of matrices*, pp. 45–8 (Addison–Wesley, 1952).

for all $x \in F$, an assertion substantiated by Lemma 6.14. The fact that the β_i are not all 0 shows that the image of F and of F/K is an infinite subgroup of C. Therefore F/K, and so G, are infinite. □

We might now ask if a presentation with negative deficiency necessarily defines a finite group. This is by no means the case.

Example 8.17. Consider $\mathrm{gp}\{x, y : x^3 = y^3 = (xy)^3 = 1\}$. This is an infinite group, for it may be verified that if

$$a = ...(012)(345)...,$$
$$b = ...(123)(456)...,$$

then $a^3 = b^3 = (ab)^3 = 1$ while $\mathrm{gp}\{a, b\}$ is infinite; for ab^2 has infinite order. On the other hand the given presentation clearly has deficiency -1. It is possible and easy to show that none of the given relations can be omitted. For instance, $\mathrm{gp}\{x, y : x^3 = y^3 = 1\}$ does not have $(xy)^3 = 1$, as we may see by taking $x = a$, $y = b^2$. Similar arguments show that the other relations are indispensable.

We have hinted that there are difficulties in any theory of finitely presented groups because of a lack of general methods and general results. This is not wholly the fault of mathematicians. Many of the fundamental difficulties arise from logical questions that are too technical to explain here. It will suffice to say that general results that are obtainable tend to be negative, and this is exemplified by the fact that the way to decide in practice whether a given finitely presented group is finite, or even trivial, is often by *ad hoc* calculation rather than by application of a theory. On the whole the finitely presented groups are a far less tractable class than (say) the free groups. If we construct groups by merely writing them down we must not be surprised to meet difficulties later on.

In conclusion we take another glance at the free groups, and prove a result which can be reached without the introduction of concepts too sophisticated for this book.

DEFINITION. Let $\mathscr{P}$ be a property which groups may possess. The group G has the property $\mathscr{P}$ *residually* if for each element x

in G with $x \neq 1$ there exists a normal subgroup N_x (depending on x) such that $x \notin N_x$ and G/N_x has the property $\mathscr{P}$.

THEOREM 8.18. *Every free group is residually finite.*

Proof. We have to prove that if x is an element of the free group F, with $x \neq 1$, then there is a normal subgroup N_x of F such that $x \notin N_x$ and F/N_x is finite. Let F have free basis $X = \{x_\lambda : \lambda \in \Lambda\}$ and let the reduced form of x be $x_{\lambda_1}^{\epsilon_1} \ldots x_{\lambda_n}^{\epsilon_n}$, with each $\epsilon_i = \pm 1$. Thus $n \geqslant 1$, and $\lambda_1, \ldots, \lambda_n$ need not all be distinct.

Let S_{n+1} denote, as usual, the symmetric group on $\{1, \ldots, n+1\}$. We shall define a mapping ϕ from F into S_{n+1} such that $x\phi \neq 1$; thus if K is the kernel of ϕ then F/K is finite, $x \notin K$, and the theorem will be proved. In fact $x\phi$ is to be a permutation mapping the symbol 1 into the symbol $n+1$. To arrange this we ask that $x_{\lambda_i}\phi = \sigma_{\lambda_i}$ where $\sigma_{\lambda_i}^{\epsilon_i}$ maps i into $i+1$, while $x_\lambda \phi = 1$ if $\lambda \notin \{\lambda_1, \ldots, \lambda_n\}$. Thus σ_{λ_i} is to map i into $i+1$ if $\epsilon_i = 1$, and $i+1$ into i if $\epsilon_i = -1$. Since F is free this mapping ϕ defined on the free basis X would determine a homomorphism from F into S_{n+1}, by Theorem 8.04.

However, we have not yet completely specified the σ_{λ_i}. Note that our requirements so far are not contradictory. If σ_{λ_i} maps i into both $i-1$ and $i+1$ then $\lambda_{i-1} = \lambda_i$, $\epsilon_{i-1} = -1$, $\epsilon_i = +1$, which is impossible because we are using the reduced form of x. Similarly if σ_{λ_i} maps both $i-1$ and $i+1$ into i then $\lambda_{i-1} = \lambda_i$, $\epsilon_{i-1} = +1$, $\epsilon_i = -1$, again impossible. Thus σ_{λ_i} is one–one, as far as it is defined, and we may complete its definition as a permutation of $n+1$ symbols in an arbitrary way.

Take $N_x = K$, and the theorem is proved. □

COROLLARY 8.19. *The intersection of all subgroups of finite index in any free group is* 1.

Proof. Let x lie in every subgroup having finite index in the free group F. If $x \neq 1$ then $x \notin N_x$, by the theorem, N_x being a certain subgroup of finite index. This contradiction shows that $x = 1$. □

The importance of residual properties is connected with the following result.

THEOREM 8.20. *If the group G has the property $\mathscr{P}$ residually then G is isomorphic to a subgroup of the cartesian product of a suitable set of groups each having the property $\mathscr{P}$.*

Proof. There are normal subgroups $\{N_x : x \in G,\ x \neq 1\}$ such that $x \notin N_x$ and G/N_x has $\mathscr{P}$. Let $N = \bigcap_{1 \neq x \in G} N_x$. Then $N \trianglelefteq G$ by Theorem 6.06. If $y \in N$ and $y \neq 1$ then $y \notin N_y$ and so $y \notin N$, a contradiction (as in Corollary 8.19). Hence $y = 1$ if $y \in N$, that is $N = 1$. By Theorem 6.32 on cartesian products $G = G/N$ is isomorphic to a subgroup of $\prod^{C}_{1 \neq x \in G} (G/N_x)$, as required. □

COROLLARY 8.21. *Every free group is isomorphic to a subgroup of the cartesian product of a suitable set of finite groups.*

Proof. Use Theorems 8.18 and 8.20. □

Problems

1. Prove the following statements about the elements a, b in an arbitrary free group:
   (i) If $a^n = b^n$ with $n > 0$ then $a = b$.
   (ii) If $a^m b^n = b^n a^m$ with $mn \neq 0$ then $ab = ba$. (Hint: use (i).)
   (iii) If $ab = ba$ then $\mathrm{gp}\{a, b\}$ is cyclic.
   (iv) If the equation $x^n = a$ has a solution x for every positive integer n then $a = 1$.

2. The group G has a normal subgroup N such that G/N is a free group. Show that G contains a subgroup F such that F is a free group, $G = NF$, and $N \cap F = 1$.

3. Let A, B denote the matrices

$$\begin{pmatrix} 1 & 1 \\ 0 & 1 \end{pmatrix}, \qquad \begin{pmatrix} 1 & 0 \\ t & 1 \end{pmatrix},$$

t being a real number which satisfies no polynomial equation with rational coefficients (which are understood to be not all zero). Prove that $\mathrm{gp}\{A, B\}$ (a multiplicative group is intended) is free and that $\{A, B\}$ is a free basis.

4. Prove that every abelian group is isomorphic to some factor group of a suitable free abelian group.

5. A certain free group F has $\{a, b\}$ for a free basis. Show that $\{ab, b\}$ is another free basis, and deduce that $\{ab^n, b\}$ is a free basis for all integers n. Which of the following are free bases for F?
(i) $\{ab, bab\}$;
(ii) $\{ab, b^{-1}ab\}$;
(iii) $\{ab, ba\}$;
(iv) $\{ab, b^2, a^2\}$;
(v) $\{a, a^{-1}b^{-1}ab\}$.

6. Find an example of a free group containing elements a, b, c, d such that $[a, b] = [c, d] \neq 1$ while $\text{gp}\{a, b\} \neq \text{gp}\{c, d\}$.

7. Let F be the free group with free basis $\{x, y\}$. Find elements a, b in F such that the length of $[a, b]$ is positive but less than the length of a and of b (lengths are referred to the given free basis).

8. In a certain free group $w_1, \ldots, w_n$ (where $n > 0$) are non-trivial elements in reduced form, for which $w_1 \ldots w_n = 1$. Prove that, for some i, all the factors in w_i are cancelled when $w_{i-1} w_i w_{i+1}$ is written in reduced form (w_0 and w_{n+1} being taken to be 1).

***9. (i) Prove that if m, n are non-zero integers and if α is a complex number with $|\alpha| \geqslant 2$ then $|2mn\alpha| \geqslant |m\alpha| + |2n|$; and that if in addition $|y_1| \geqslant |y_2|$ for complex y_1, y_2 then

$$|(1+2mn\alpha)y_1 + 2ny_2| \geqslant |m\alpha y_1 + y_2|.$$

(ii) Let $G(\alpha, \beta)$ be the multiplicative group generated by the matrices

$$A = \begin{pmatrix} 1 & \alpha \\ 0 & 1 \end{pmatrix}, \qquad B = \begin{pmatrix} 1 & 0 \\ \beta & 1 \end{pmatrix},$$

where α, β are complex. By considering $X^{-1}AX$ and $X^{-1}BX$, where

$$X = \begin{pmatrix} 1 & 0 \\ 0 & \gamma \end{pmatrix},$$

prove that if $\gamma \neq 0$ then $G(\alpha, \beta) \cong G(\alpha\gamma, \beta/\gamma)$.

(iii) By considering the second rows of Y and YA^mB^n, where

$$Y = \begin{pmatrix} x_1 & x_2 \\ y_1 & y_2 \end{pmatrix}$$

and m, n are non-zero, prove that $G(\alpha, \beta)$ is free if $|\alpha\beta| \geqslant 4$.

10. Let $\mathscr{G} = \text{gp}\{A, B\}$ where the matrices A, B are

$$\begin{pmatrix} 1 & 1 \\ 0 & 1 \end{pmatrix}, \qquad \begin{pmatrix} 2 & 0 \\ 0 & 1 \end{pmatrix}$$

respectively (as in Problem 11 of Chapter 3). Prove (using the result of that problem, or otherwise) that

$$\mathscr{G} \cong \mathrm{gp}\{a, b : bab^{-1} = a^2\}.$$

11. Prove that $\mathrm{gp}\{a, b\}$ is infinite if

$$a = \ldots(1,\ldots, n)(n+1,\ldots, 2n)\ldots(in+1,\ldots, (i+1)n)\ldots,$$
$$b = \ldots(2,\ldots, n+1)(n+2,\ldots, 2n+1)\ldots(in+2,\ldots. (i+1)n+1)\ldots,$$

and $n \geqslant 2$.

Using this result prove that

$$\mathrm{gp}\{a, b : a^{-1}b^{\alpha}a = b^{\beta},\ b^{-1}a^{\alpha}b = a^{\beta}\}$$

is infinite if α, β have a common factor greater than 1.

12. The elements a, b of the group G are such that

$$b^{-1}a^{\alpha}b = a^{\beta}, \qquad a^{-1}b^{\gamma}a = b^{\delta}$$

where α, γ are positive. By finding two expressions for $a^{\beta\gamma}$, or otherwise, prove that $b^{\gamma-\delta}$ commutes with $a^{\alpha\gamma}$ and with $a^{\beta\gamma}$. Deduce that if α and β are coprime then $b^{\gamma-\delta}$ commutes with a, and that $b^{(\gamma-\delta)^2} = 1$.

Prove that $\mathrm{gp}\{a, b : b^{-1}a^{\alpha}b = a^{\alpha+1}, a^{-1}b^{\gamma}a = b^{\gamma+1}\}$ has order 1; and that $\mathrm{gp}\{a, b : b^{-1}a^3b = a^5, a^{-1}b^7a = b^9\}$ is the quaternion group.

13. What is the order of the group presented by generators a, b and defining relations

$$a^3b^{-1}a^{-2}ba^{-1}b^2ab^{-3}a^{-1}b^2ab^{-3}b^{-1}a^2ba^{-3}b^3a^{-1}b^{-2}ab^3a^{-1}b^{-2}ab^3a^{-1}b^{-2}a = 1,$$
$$b^3a^{-1}b^{-2}ab^{-1}a^2ba^{-3}b^{-1}a^2ba^{-3}a^{-1}b^2ab^{-3}a^3b^{-1}a^{-2}ba^3b^{-1}a^{-2}ba^3b^{-1}a^{-2}b = 1\,?$$

*14. (i) Prove that if a, b are elements of some group for which there exist integers γ, δ giving $a^{\gamma}ba^{\delta}b = 1$ then $a^{\gamma-\delta}$ commutes with b.

(ii) Let $G = \mathrm{gp}\{a, b : ba^{\alpha}b^2 = a^2,\ ba^{\beta}b = a\}$. Prove (by considering $G/\delta(G)$ or otherwise) that G is infinite if $2\alpha - 3\beta = 1$ and, in particular, if $\alpha = \beta = -1$.

(iii) Assume now that the integers α, β associated with G are arbitrary. Prove (by means of (i) or otherwise) that $a^{\beta+1}$ and $a^{2\alpha+2}$ both commute with b.

(iv) Prove that if $c = a^{\alpha-\beta-1}b$ then $aca^{-1} = c^2$.

(v) Deduce that if $2\alpha - 3\beta \neq 1$, then G is finite.

(vi) Show that (iii) implies that G is abelian if $\beta + 1$ and $2\alpha + 2$ are coprime.

*15. Which of the following statements are correct? Prove each answer you give.

(i) If every subgroup generated by three elements of G is free then G itself is free.

(ii) No free group is a (non-trivial) direct product.
(iii) Every subgroup of a residually finite group is residually finite.
(iv) Every factor group of a residually finite group is residually finite.
(v) If the free group F_m of rank m is generated by a set of n elements, where $n > m$, then a subset of m of these elements will provide a free basis for F_m.

***16. Let F be a free group with free basis $X = \{x_\lambda : \lambda \in \Lambda\}$. Prove that F is residually a finite p-group by means of the following steps:

(i) Take an arbitrary non-trivial reduced word $x_{\lambda_1}^{m_1} \dots x_{\lambda_n}^{m_n}$ in F with each m_i non-zero and $\lambda_i \neq \lambda_{i+1}$ for $1 \leqslant i < n$. Show that there is a power p^α of p for which $m_1 \dots m_n \not\equiv 0$ modulo p^α.

(ii) Let $\mathscr{U}$ be the set of all $(n+1)\times(n+1)$ matrices with integers modulo p^α for entries, each element on the diagonal being 1 and each element below the diagonal being 0. Prove that $\mathscr{U}$, and all its subgroups, form (under multiplication) finite p-groups.

(iii) Let I be the unit matrix in $\mathscr{U}$, and E_{jk} the matrix whose only non-zero entry is 1 in row j, column k. Prove that $E_{j_1k_1}E_{j_2k_2} = 0$ if $k_1 \neq j_2$, and that $E_{j_1k_1}E_{j_2k_2} = E_{j_1k_2}$ if $k_1 = j_2$. Deduce that $(I+E_{jk})^m = I+mE_{jk}$, if $j \neq k$.

(iv) Define $X_\mu = \prod_{\lambda_i=\mu} (I+E_{i,i+1})$ with $X_\mu = I$ if no λ_i equals μ, and show that the entry in row 1, column $n+1$ of $X_{\lambda_1}^{m_1} \dots X_{\lambda_n}^{m_n}$ is $m_1 \dots m_n$.

(v) Construct a homomorphism from F into $\mathscr{U}$ under which the image of $x_{\lambda_1}^{m_1} \dots x_{\lambda_n}^{m_n}$ is non-trivial.

9

Nilpotent groups

THE theorem that states that a finite group with p^α dividing its order has a subgroup of order p^α indicates that the study of the finite p-groups is basic to any theory of finite groups. Many of the properties of finite p-groups are shared by a wider class of groups, which might be expected to be the general (finite or infinite) p-groups, but which are in fact the nilpotent groups defined below. These are much more agreeable to work with than the infinite p-groups, and seem to be the 'correct' generalization of the finite p-groups.

In connection with the series about to be introduced, we again state that all series considered in this book are *finite*.

DEFINITION. A *central series* in the group G is a normal series

$$G = G_0 \trianglerighteq G_1 \trianglerighteq \dots \trianglerighteq G_{r-1} \trianglerighteq G_r = 1$$

such that

(i) G_i is a normal subgroup of G, for $0 \leqslant i \leqslant r$; and
(ii) G_{i-1}/G_i lies in the centre of G/G_i, for $1 \leqslant i \leqslant r$.

(Note that (i) follows from (ii) and the fact that the central subgroup G_{r-1} is normal in G.)

DEFINITION. A group is *nilpotent* if it has a central series. A group with a central series of length r is said to be *nilpotent of class r*.

Note that, as we have defined it, the class is not an invariant of a nilpotent group—the length of a central series may always be increased by inserting repeated terms. The *least* value of r associated with the group is, of course, an invariant.

Abelian groups are nilpotent, being of class 1. Other examples of nilpotent groups are the groups G in which $\delta(G) \leqslant \zeta(G)$, for we may take $r = 2$, $G_1 = \zeta(G)$ and then $G/\zeta(G)$ is abelian, in accordance with Theorem 4.21. Further examples will appear in the natural course of our discussion.

It will be found, on the other hand, that the symmetric group S_3 has no central series. The term G_{r-1} in the above standard series must be central in G, and S_3 has trivial centre.

We need some minor work in order to apply the above definitions with greater ease. The first item is essentially a matter of notation. If A, B are subgroups of the group G then $[A, B]$ will denote $\mathrm{gp}\{[a, b]: a \in A,\ b \in B\}$. We make two observations: if $A \trianglelefteq G$ then $[A, B] \leqslant A$, and if $A \leqslant A_1$, $B \leqslant B_1$ then $[A, B] \leqslant [A_1, B_1]$.

LEMMA 9.01. *Let*

$$G = G_0 \geqslant G_1 \geqslant \ldots \geqslant G_{r-1} \geqslant G_r = 1$$

be a series of subgroups each normal in G. This is a central series if and only if $[G_{i-1}, G] \leqslant G_i$ for $1 \leqslant i \leqslant r$.

Proof. The central property for the series just mentioned is equivalent to $G_{i-1}/G_i \leqslant \zeta(G/G_i)$ for each i. This in turn is equivalent to $[yG_i, xG_i] = G_i$ or $[y, x]G_i = G_i$ or $[y, x] \in G_i$ for each i and for all x in G, y in G_{i-1}. But this is equivalent to $[G_{i-1}, G] \leqslant G_i$. □

THEOREM 9.02. *All subgroups and factor groups of the nilpotent group G are nilpotent. If G is of class r then every subgroup and factor group has class r.*

Proof. Let G have the standard central series which we described in Lemma 9.01 and let S be a subgroup of G. Consider the series

$$S = S_0 \geqslant S_1 \geqslant \ldots \geqslant S_{r-1} \geqslant S_r = 1$$

in S, where S_i is defined to be $S \cap G_i$ for $0 \leqslant i \leqslant r$. We have $S_i \trianglelefteq S$ by Corollary 6.02, because $G_i \trianglelefteq G$. By Lemma 9.01, the series in S will be central if $[S_{i-1}, S] \leqslant S_i$ for $1 \leqslant i \leqslant r$. But $S_{i-1} \leqslant G_{i-1}$, $S \leqslant G$, and so $[S_{i-1}, S] \leqslant [G_{i-1}, G]$. The series for G was central, so by the lemma $[G_{i-1}, G] \leqslant G_i$, and we have $[S_{i-1}, S] \leqslant G_i$. Clearly $[S_{i-1}, S] \leqslant S$, and we thus have $[S_{i-1}, S] \leqslant S \cap G_i = S_i$, as desired.

Next let N be any normal subgroup of G, and consider the following series in G/N:

$$G/N = G_0 N/N \geqslant G_1 N/N \geqslant \ldots \geqslant G_{r-1} N/N \geqslant G_r N/N = N.$$

That this is a series, and that $G_i N/N \trianglelefteq G/N$, require only elementary verification. Typical elements of G/N, $G_{i-1} N/N$ are xN, yN respectively where $x \in G$, $y \in G_{i-1}$; and

$$[yN, xN] = [y, x]N \in G_i N/N$$

because $[y, x] \in G_i$. Hence $[G_{i-1} N/N,\ G/N] \leqslant G_i N/N$, and the theorem is proved. □

Note that S_n cannot be nilpotent for $n \geqslant 3$, by this theorem. For if S_n were nilpotent then its subgroup S_3 would also be nilpotent.

THEOREM 9.03. *The direct product of a finite set of nilpotent groups is nilpotent.*

Proof. It will suffice to prove this theorem for the direct product of *two* nilpotent groups; once that is done, an easy inductive argument (which is omitted) would complete the proof. We lose no further generality in supposing that we are given nilpotent groups H, K with central series of the same length r:

$$H = H_0 \geqslant H_1 \geqslant \ldots \geqslant H_{r-1} \geqslant H_r = 1,$$

$$K = K_0 \geqslant K_1 \geqslant \ldots \geqslant K_{r-1} \geqslant K_r = 1;$$

for if not we may lengthen the shorter series by inserting repeated members. We shall prove that $G = H \times K$ has a central series of length r, given by $G_i = H_i \times K_i$ for $0 \leqslant i \leqslant r$.

It is not hard to see that $G_i \trianglelefteq G$, so we omit formal proof of this fact. We have yet to show that $[G_{i-1}, G] \leqslant G_i$ for $1 \leqslant i \leqslant r$, if we are to use Lemma 9.01 in this proof. But an easy computation gives

$$[H_{i-1}\times K_{i-1}, H\times K] = [H_{i-1}, H]\times[K_{i-1}, K] \leqslant H_i\times K_i$$

for $1 \leqslant i \leqslant r$; that is, $[G_{i-1}, G] \leqslant G_i$. The proof is complete. □

Notice that in the proof just given we are identifying direct factors with subgroups of the direct product in the usual way, which was explained in Theorem 4.38.

It is not true that the direct product of an arbitrary set of nilpotent groups is nilpotent. To see this one forms $\prod^D_{n>0} G_n$ where $G_1, G_2,..., G_n,...$ are groups such that G_n is nilpotent but of no class less than n. Such groups do exist (see Problems 1 and 3 of this chapter). If the direct product were nilpotent then it would have class c, say, and so each G_n as a subgroup of G would also have class c (by Theorem 9.02), a contradiction. Of course $\prod^C_{n>0} G_n$ certainly cannot be nilpotent, either.

THEOREM 9.04. *Any finite p-group is nilpotent.*

Proof. Let G be a finite p-group with order p^n. (We use Corollary 7.06 without explicit reference from now on.) The proof goes by induction on n. The cases $n = 0, 1$ are trivially easy, so we take $n > 1$. Theorem 4.29 ensures that $\zeta(G) > 1$, and this is the key to the present proof because the inductive hypothesis may be applied to $G/\zeta(G)$. Since this group has order strictly less than $|G|$, it has a central series:

$$G/\zeta(G) = G_0/\zeta(G) \geqslant G_1/\zeta(G) \geqslant ... \geqslant G_{r-1}/\zeta(G) \geqslant G_r/\zeta(G) = \zeta(G),$$

the normal subgroups G_i of G each containing $\zeta(G)$. We have $[G_{i-1}/\zeta(G), G/\zeta(G)] \leqslant G_i/\zeta(G)$ for $1 \leqslant i \leqslant r$, by Lemma 9.01. Therefore we have $[y\zeta(G), x\zeta(G)] \in G_i/\zeta(G)$ for all $x \in G, y \in G_{i-1}$. Thus $[y, x]\zeta(G) \in G_i/\zeta(G)$, and $[y, x] \in G_i$. Now consider the series

$$G = G_0 \geqslant G_1 \geqslant ... \geqslant G_{r-1} \geqslant G_r \geqslant G_{r+1} = 1$$

for G. We have $G_i \trianglelefteq G$ and $[G_{i-1}, G] \leqslant G_i$ for $1 \leqslant i \leqslant r$, as above; further, $[G_r, G] \leqslant G_{r+1}$ since $G_r = \zeta(G)$ and $G_{r+1} = 1$. Thus we have a central series for G, and G is nilpotent. □

In fact the proof just given can be made to yield a little more: if the group G is such that $G/\zeta(G)$ has class r then G itself has class $r+1$. The reader should prove this for himself.

The next result is of fundamental importance.

THEOREM 9.05. *If the nilpotent group G has a proper subgroup H then H is a proper subgroup of its normalizer.*

Proof. Take a central series for G:

$$G = G_0 \geqslant G_1 \geqslant \ldots \geqslant G_{r-1} \geqslant G_r = 1.$$

We have $[G_{i-1}, G] \leqslant G_i$ for $1 \leqslant i \leqslant r$, by Lemma 9.01. Suppose that $G_k \leqslant H$ while $G_{k-1} \not\leqslant H$. Such a value of k exists because $1 = G_r \leqslant H$ and $G = G_0 \not\leqslant H$, H being a proper subgroup of G. Then

$$[G_{k-1}, G] \leqslant G_k \leqslant H,$$

and so $[G_{k-1}, H] \leqslant H$. In other terms, G_{k-1} normalizes H. By the choice of k there is an element of G_{k-1} which does not lie in H, and it follows that $N(H) > H$. □

COROLLARY 9.06. *If the nilpotent group G has a maximal subgroup M then $M \trianglelefteq G$ and G/M has prime order.*

Proof. We have $M < G$ and so $N(M) = G$, by the theorem, since $M \leqslant H \leqslant G$ implies $M = H$ or $H = G$. Therefore $M \trianglelefteq G$. By Corollary 6.05 G/M has no proper subgroup, and this easily implies that G/M has prime order. □

Note carefully that we do not assert that a nilpotent group must have a maximal subgroup. In fact this is not the case even for abelian groups. The infinite abelian group Z_p^∞ of Example 1.09 has every proper subgroup finite cyclic, as the reader may care to prove, and this ensures that no maximal subgroup exists.

COROLLARY 9.07. *If the nilpotent group G has a Sylow π-subgroup G_π, for some set π of primes, then G_π is normal in G.*

Proof. If G_π is a Sylow π-subgroup then $N(G_\pi) = N(N(G_\pi))$ by Corollary 7.21. On applying Theorem 9.05 to the subgroup $N(G_\pi)$ of G, we conclude that $G = N(G_\pi)$. That is, $G_\pi \trianglelefteq G$. □

We have not asserted that any Sylow π-subgroup exists. But we shall soon prove that G_π does exist, by a method that avoids set-theoretic difficulties.

At this point it is possible to give much information about finite nilpotent groups and, in particular, to characterize them in terms of finite p-groups. The following theorem generalizes some familiar facts about abelian groups.

THEOREM 9.08. *The following conditions on the finite group G are equivalent.*

(i) *G is nilpotent.*
(ii) *All maximal subgroups of G are normal.*
(iii) *Any Sylow p-subgroup G_p is normal (p is an arbitrary prime).*
(iv) *Elements of coprime orders in G commute.*
(v) *G is the direct product of its Sylow p-subgroups.*

Proof. (i) implies (ii) by Corollary 9.06.

Since G is finite, maximal subgroups exist; and so (ii) implies (iii) by Corollary 7.30.

To prove that (iii) implies (iv), we first observe that elements of orders p^α, q^β (where p, q are distinct primes) lie in normal Sylow subgroups G_p, G_q respectively. Since their commutator lies in $G_p \cap G_q = 1$, they commute.

Next suppose that a, b are elements of orders $p_1^{\alpha_1} \dots p_u^{\alpha_u}$, $q_1^{\beta_1} \dots q_v^{\beta_v}$ respectively, where the primes $p_1, \dots, p_u, q_1, \dots, q_v$ are distinct. A routine application of the Euclidean algorithm shows that $a = a_1 \dots a_u$, $b = b_1 \dots b_v$ where a_i, b_j have orders $p_i^{\alpha_i}$, $q_j^{\beta_j}$ respectively (compare Theorem 5.01). Since each a_i commutes with each b_j, as above, we find that a commutes with b. Therefore (iii) implies (iv).

We deduce (v) from (iv) as follows. Take one Sylow p-subgroup G_p for each prime p dividing $|G|$. Then (iv) asserts that G_p commutes with G_q element by element, provided $p \neq q$. Suppose that $|G| = p_1^{\alpha_1} \dots p_u^{\alpha_u}$, so that $|G_{p_i}| = p_i^{\alpha_i}$; it follows on comparison of orders that $G = G_{p_1} \dots G_{p_u}$ (note the use of

Theorem 3.20 here). It should now be clear that

$$G = G_{p_1} \times \ldots \times G_{p_u}$$

by virtue of the direct product criterion of Theorem 4.38. Therefore (iv) implies (v).

To finish off the proof, we show that (v) implies (i). This, however, is an easy consequence of Theorems 9.04 and 9.03. □

Some results below are best reached by means of special central series. We therefore define inductively subgroups $\zeta_i(G)$ and $\gamma_i(G)$ of the arbitrary group G. Let $\zeta_0(G) = 1$, and let $\zeta_i(G)$ be that subgroup of G for which

$$\zeta_i(G)/\zeta_{i-1}(G) = \zeta\{G/\zeta_{i-1}(G)\}$$

for each $i \geqslant 1$. Let $\gamma_1(G) = G$, and let $\gamma_i(G) = [\gamma_{i-1}(G), G]$ for each $i > 1$. Note that $\zeta_1(G) = \zeta(G)$ and $\gamma_2(G) = \delta(G)$.

DEFINITION. If $\zeta_s(G) = G$ for some integer s then G is said to have the *upper central series*

$$1 = \zeta_0(G) \leqslant \zeta_1(G) \leqslant \ldots \leqslant \zeta_{s-1}(G) \leqslant \zeta_s(G) = G.$$

DEFINITION. If $\gamma_{t+1}(G) = 1$ for some integer t then G is said to have the *lower central series*

$$G = \gamma_1(G) \geqslant \gamma_2(G) \geqslant \ldots \geqslant \gamma_t(G) \geqslant \gamma_{t+1}(G) = 1.$$

We ought to show that these are indeed central series as this term was defined earlier. The normal property of each $\zeta_i(G)$ should be clear, and our earlier definition ensures that the $\zeta_i(G)$ give a central series. We prove that $\gamma_i(G) \trianglelefteq G$ by induction on i, the case $i = 1$ being trivial. If $x \in G$, $y \in \gamma_{i-1}(G)$ then

$$[y, x]^g = [y^g, x^g] \in [\gamma_{i-1}(G), G]$$

for all $g \in G$; it follows that $\gamma_i(G)^g \leqslant \gamma_i(G)$, which suffices. That the $\gamma_i(G)$ give a central series follows from Lemma 9.01.

It is not hard to show that $\zeta_i(G)$ and $\gamma_i(G)$ are characteristic subgroups of G.

Example 9.09. Let $G = Q_8 \times C$ where

$$Q_8 = \text{gp}\{a, b : a^2 = b^2 = (ab)^2\},$$
$$C = \text{gp}\{c : c^2 = 1\}.$$

It is easy to calculate that

$$\gamma_2(G) = \text{gp}\{a^2\}, \qquad \gamma_3(G) = 1,$$
$$\zeta_1(G) = \text{gp}\{a^2, c\}, \quad \zeta_2(G) = G,$$

in a lax but clear notation. Here is a case in which the upper and lower central series exist and have the same length, but possess differing terms.

THEOREM 9.10. *If G has a central series*

$$G = G_0 \geqslant G_1 \geqslant \ldots \geqslant G_{r-1} \geqslant G_r = 1$$

then $G_{r-i} \leqslant \zeta_i(G)$ and $G_i \geqslant \gamma_{i+1}(G)$ for $0 \leqslant i \leqslant r$.

Proof. When $i = 0$ both results are trivial. We use induction on i, taking $1 \leqslant i < r$.

In the former case we suppose inductively that

$$G_{r-i+1} \leqslant \zeta_{i-1}(G).$$

Take an element x in G_{r-i}, with the aim of showing that x lies in $\zeta_i(G)$; take also an arbitrary element y in G. We then have $[x, y] \in G_{r-i+1}$, by Lemma 9.01, and now the inductive hypothesis implies that $[x, y] \in \zeta_{i-1}(G)$. This means that the elements $\zeta_{i-1}(G)x$ and $\zeta_{i-1}(G)y$ of the group $G/\zeta_{i-1}(G)$ commute. But y was arbitrary in G, which shows that $\zeta_{i-1}(G)y$ is arbitrary in $G/\zeta_{i-1}(G)$ and hence that $\zeta_{i-1}(G)x$ is a central element of $G/\zeta_{i-1}(G)$. Now the centre of $G/\zeta_{i-1}(G)$ is $\zeta_i(G)/\zeta_{i-1}(G)$, by definition of the upper central series. It follows that $x \in \zeta_i(G)$ and, of course, that $G_{r-i} \leqslant \zeta_i(G)$, which is the result we require.

For the second case we make the inductive assumption that $G_{i-1} \geqslant \gamma_i(G)$. Then

$$\gamma_{i+1}(G) = [\gamma_i(G), G] \leqslant [G_{i-1}, G] \leqslant G_i,$$

the last inclusion following from Lemma 9.01. Thus we have the desired result $G_i \geqslant \gamma_{i+1}(G)$. □

Corollary 9.11. *If a group is nilpotent then its upper and lower central series have the same length, and this is the least length for any central series.*

Proof. The theorem clearly implies that the given series of length r is at least as long as the upper and lower central series. Since the given series was an arbitrary central series we may thus compare the lengths of the upper and lower central series. They are equal. □

We turn to a closer study of the lower central series.

Lemma 9.12. *If x, y, z are any elements of an arbitrary group then*

(i) $[xy, z] = [x, z]^y[y, z]$;
(ii) $[x, yz] = [x, z][x, y]^z$.

Proof. We recall that u^v means $v^{-1}uv$. We have

$$(xy)^z = x^z y^z,$$

$$[xy, z] = (xy)^{-1}(xy)^z = y^{-1}x^{-1}x^z y^z = (x^{-1}x^z)^y(y^{-1}y^z)$$
$$= [x, z]^y[y, z],$$

and this gives (i). Because $[u, v]^{-1} = [v, u]$ inversion of both sides of (i) gives
$$[z, xy] = [z, y][z, x]^y,$$

which is (ii) with suitable relabelling. □

Corollary 9.13. *In the same notation*

(iii) $[x^{-1}, y] = [x, y]^{-x^{-1}}$;
(iv) $[x, y^{-1}] = [x, y]^{-y^{-1}}$.

Proof. Here u^{-v} means $(u^{-1})^v$. Put $y = x^{-1}$ in (i) and obtain

$$1 = [x, z]^{x^{-1}}[x^{-1}, z].$$

Then (iii) follows, and the proof of (iv) is similar. □

DEFINITION. The *simple commutator* $[x_1,\ldots, x_n]$ where $n \geqslant 2$ and $x_1,\ldots, x_n$ are elements of a fixed group G is defined inductively by $[x_1, x_2] = x_1^{-1}x_2^{-1}x_1x_2$ and by $[x_1,\ldots, x_n] = [[x_1,\ldots, x_{n-1}], x_n]$ for all $n > 2$.

THEOREM 9.14. *If* $G = \text{gp}\{a_\lambda : \lambda \in \Lambda\}$ *then, for each* $i > 1$, $\gamma_i(G)/\gamma_{i+1}(G)$ *is generated by the set of all elements* $[b_1,\ldots, b_i]\gamma_{i+1}(G)$, *where each* b_j *is chosen arbitrarily from* $\{a_\lambda : \lambda \in \Lambda\}$.

Proof. Note that the theorem would be valid for $i = 1$ if we defined $[b_1,\ldots, b_i]$ to be b_1 when $i = 1$. We shall adopt this convention and use the case $i = 1$ as the basis of an inductive proof. (It is, of course, feasible to prove first the case $i = 2$ and then the inductive step, but they are essentially the same proof and our device avoids such repetition.) Let $i > 1$ and suppose that $\gamma_{i-1}(G)/\gamma_i(G)$ has a generating set of the required form.

The definition of $\gamma_i(G)$ implies that

$$\gamma_i(G) = \text{gp}\{[y, x] : y \in \gamma_{i-1}(G),\ x \in G\}$$

for all $i > 1$. Therefore the elements $[y, x]\gamma_{i+1}(G)$ generate $\gamma_i(G)/\gamma_{i+1}(G)$. Now the arbitrary element y of $\gamma_{i-1}(G)$ has the form cz where $c = c_1 \ldots c_r$, each c_j being a commutator $[b_1,\ldots, b_{i-1}]$ or the inverse of such a commutator, and $z \in \gamma_i(G)$; this is the inductive assumption. But

$$\begin{aligned}[cz, x] &= [c, x]^z[z, x]\\ &= [c, x][c, x, z][z, x],\end{aligned}$$

by formula (i) of Lemma 9.12, and the fact that $u^v = u[u, v]$, respectively. Because $[c, x, z]$ and $[z, x]$ lie in $\gamma_{i+1}(G)$ we have

$$[cz, x]\gamma_{i+1}(G) = [c, x]\gamma_{i+1}(G),$$

and it follows that the elements $[c, x]\gamma_{i+1}(G)$ generate $\gamma_i(G)/\gamma_{i+1}(G)$.

Consider $[c, x]\gamma_{i+1}(G)$. A similar application of Lemma 9.12 gives

$$[c, x]\gamma_{i+1}(G) = [c_1, x] \dots [c_r, x]\gamma_{i+1}(G).$$

If it happens that it is c_j^{-1} which has the form $[b_1, \dots, b_{i-1}]$, then we apply formula (iii) of Corollary 9.13:

$$\begin{aligned}[c_j, x] &= [c_j^{-1}, x]^{-c_j} \\ &= [c_j^{-1}, x]^{-1}[[c_j^{-1}, x]^{-1}, c_j].\end{aligned}$$

(Again we also use the fact that $u^v = u[u, v]$.) Now $[[c_j^{-1}, x]^{-1}, c_j] \in \gamma_{i+1}(G)$, and so

$$[c_j, x]\gamma_{i+1}(G) = [c_j^{-1}, x]^{-1}\gamma_{i+1}(G).$$

We may conclude that the elements $[b_1, \dots, b_{i-1}, x]\gamma_{i+1}(G)$ generate $\gamma_i(G)/\gamma_{i+1}(G)$.

Here x is still an arbitrary element of G. We put $x = x_1 \dots x_s$ where each x_j equals some a_λ or a_λ^{-1}, and expand the commutator $[b_1, \dots, b_{i-1}, x]$ using (ii) and (iv). The details follow, with $b_0 = [b_1, \dots, b_{i-1}]$:

$$[b_0, x_1 \dots x_s]\gamma_{i+1}(G) = [b_0, x_1] \dots [b_0, x_s]\gamma_{i+1}(G),$$

$$[b_0, a_\lambda^{-1}]\gamma_{i+1}(G) = [b_0, a_\lambda]^{-1}\gamma_{i+1}(G).$$

Therefore $\gamma_i(G)/\gamma_{i+1}(G)$ is generated by the elements

$$[b_1, \dots, b_{i-1}, a_\lambda]\gamma_{i+1}(G),$$

and this completes the inductive proof of the theorem. □

COROLLARY 9.15. *If the group G is finitely generated then $\gamma_i(G)/\gamma_{i+1}(G)$ is finitely generated for each $i \geqslant 1$.* □

Next we apply the above theorem to nilpotent groups.

THEOREM 9.16. *Every subgroup of a finitely generated nilpotent group is finitely generated.*

Proof. Let H be a subgroup of the finitely generated group G and let G have lower central series of length s. Theorem 9.14 shows that $\gamma_i(G)/\gamma_{i+1}(G)$ is finitely generated for $1 \leqslant i \leqslant s$. It is easy to verify again that the series

$$H = H_1 \geqslant H_2 \geqslant \dots \geqslant H_s \geqslant H_{s+1} = 1,$$

where $H_i = H \cap \gamma_i(G)$ for $1 \leqslant i \leqslant s+1$, is a central series (perhaps not the lower central series) for H. Further, we have

$$\begin{aligned} H_i/H_{i+1} &= \{H \cap \gamma_i(G)\}/\{H \cap \gamma_i(G) \cap \gamma_{i+1}(G)\} \\ &\cong \gamma_{i+1}(G)\{H \cap \gamma_i(G)\}/\gamma_{i+1}(G) \end{aligned}$$

by the second isomorphism theorem. Therefore H_i/H_{i+1} is isomorphic to a subgroup of $\gamma_i(G)/\gamma_{i+1}(G)$. But any subgroup of a finitely generated abelian group is finitely generated, according to Theorem 5.18. It follows that H_i/H_{i+1} is finitely generated for $1 \leqslant i \leqslant s$.

It is easy to see that if $N \trianglelefteq G$ and both N and G/N are finitely generated then G is finitely generated. An extension of this remark, which can be proved rigorously by induction, shows that each H_i/H_{i+1} being finitely generated implies that H is finitely generated. □

THEOREM 9.17. *A nilpotent group is finite if it is generated by a finite number of elements each having finite order.*

Proof. Let G be nilpotent and let G be generated by the subset $\{a_1,\ldots, a_n\}$ of elements, a_i having finite order for $1 \leqslant i \leqslant n$. The fact that G has a lower central series makes it sufficient to prove that $\gamma_i(G)/\gamma_{i+1}(G)$ is finite for each $i \geqslant 1$. This factor group is abelian and, as Corollary 9.15 indicates, finitely generated. Therefore it will suffice to produce for $\gamma_i(G)/\gamma_{i+1}(G)$ a generating set each element of which has finite order.

If we refer to Theorem 9.14 we will find that the key step was this fact: if

$$\gamma_{i-1}(G)/\gamma_i(G) = \mathrm{gp}\{b'_1\gamma_i(G),\ldots, b'_m\gamma_i(G)\}$$

(where $i > 1$) then

$$\gamma_i(G)/\gamma_{i+1}(G) = \mathrm{gp}\{[b'_j, a_k]\gamma_{i+1}(G) : 1 \leqslant j \leqslant m,\ 1 \leqslant k \leqslant n\}.$$

If $a_k^{\alpha_k} = 1$ for some positive integer α_k then it is certainly true that $[b'_j, a_k^{\alpha_k}] \in \gamma_{i+1}(G)$. The commutator $[b'_j, a_k^{\alpha_k}]$ may be expanded by means of formula (ii) in Lemma 9.12, and (because $b'_j \in \gamma_{i-1}(G)$ with the result that $[b'_j, a_k] \in \gamma_i(G)$ while $[b'_j, a_k, x] \in \gamma_{i+1}(G)$) we find that $[b'_j, a_k]^{\alpha_k}\gamma_{i+1}(G) = \gamma_{i+1}(G)$.

Therefore the element $[b_j', a_k]\gamma_{i+1}(G)$ of $\gamma_i(G)/\gamma_{i+1}(G)$ has order dividing α_k, and certainly has finite order.

Since we now have a finite set of elements generating $\gamma_i(G)/\gamma_{i+1}(G)$, each element having finite order, the theorem follows by the remarks above. □

COROLLARY 9.18. *In any nilpotent group G the elements of finite order form a normal subgroup N such that G/N is torsionfree.*

Proof. Let x, y be elements of finite order in G. By the theorem $\text{gp}\{x, y\}$ is finite, and it follows that xy^{-1} has finite order. Since 1 has finite order, the subgroup criterion (Theorem 3.06) now shows that the set of elements of finite order forms a subgroup, N say. It is clear that $N \trianglelefteq G$. If Na is an element of G/N with finite order α then $(Na)^\alpha = N$ and so $a^\alpha \in N$. Since N is periodic a^α, and so a, have finite order. Therefore $a \in N$, and $Na = N$. This shows that G/N is torsionfree. □

COROLLARY 9.19. *In any nilpotent group elements of coprime orders commute.*

Proof. We are referring, of course, only to elements of finite order. If a, b are elements of a nilpotent group with orders α, β respectively then $\text{gp}\{a, b\}$ is finite, by Theorem 9.17. Let α, β be coprime. An earlier theorem (9.08) about *finite* nilpotent groups then shows that a and b commute. □

COROLLARY 9.20. *For each prime p the nilpotent group G contains one and only one Sylow p-subgroup, which is therefore normal in G.*

Proof. Given p and G, we consider the set S of all elements of p-power order. A proof very like the one given in Corollary 9.18 (using Theorem 9.08 as well) shows that S is a subgroup. It is then clear (from the definition of S) that there is precisely one Sylow p-subgroup, namely S. Corollary 7.17 implies that $S \trianglelefteq G$. □

COROLLARY 9.21. *The periodic subgroup of the nilpotent group G is the direct product of the Sylow p-subgroups of G for all primes p.*

Proof. By the 'periodic subgroup' of G we mean that subgroup N defined in Corollary 9.18. That $S_p \leqslant N$, where S_p is the Sylow p-subgroup described in Corollary 9.20, is a trivial fact. A simple application of the direct product criterion (6.30), with the aid of Corollary 9.19, gives the result—the details are no harder than in an earlier case in which G was finite. □

Example 9.22. We construct a finitely generated nilpotent group G whose periodic subgroup is not a direct factor of G. Let $Q_8 = \text{gp}\{a, b : a^2 = b^2 = (ab)^2\}$ be the quaternion group of order 8, and let $Z = \text{gp}\{z\}$ be infinite cyclic. Our example G is the subgroup of $Q_8 \times Z$ generated by $\{x, y\}$ where $x = (a, z)$, $y = (b, 1)$. It is easy to see that the periodic subgroup P is $\text{gp}\{y, [x, y]\}$, which is abelian, and that G/P is infinite cyclic. This implies that either P is a direct factor and G is abelian, which is not the case, or P is not a direct factor.

In Chapter 11 we shall construct an abelian group whose periodic subgroup is not a direct summand (see Example 11.14).

The theory of nilpotent groups developed here can be extended greatly, but, of course, our account is restricted by the fact that only a small elementary portion is accessible to us. One general remark that we might make is that its conclusions do not apply to all p-groups, for there exist infinite p-groups which are not nilpotent. An example was indicated on page 175, and another example that even has trivial centre will be found among the problems at the end of this chapter. (See Problem 17.)

The less general theorems that follow may be found to have some appeal on the grounds of interest, elegance, or importance.

The theory of nilpotent groups has an application to finite p-groups and their sets of generating elements. Our context will now be finite groups (not necessarily p-groups), so that we may avoid difficulties associated with infinite set theory.

DEFINITION. The *Frattini subgroup* of the group G is the intersection of all the maximal subgroups of G. We adopt the notation $\Phi(G)$ for this subgroup, and the convention that $\Phi(G) = G$ if G has no maximal subgroup.

THEOREM 9.23. *The Frattini subgroup of a finite group consists of all elements which may be omitted from any generating set (for the group) in which they occur.*

Proof. Let x be an element of the finite group G and suppose that $x \notin \Phi(G)$. Then $x \notin M$ where M is a certain maximal subgroup of G. It is then clear that $G = \mathrm{gp}\{M, x\}$ while $G > \mathrm{gp}\{M\} = M$. Thus $\{M, x\}$ is a set of generators for G while $\{M\}$ is not. It follows that if the element x of G can be omitted from every generating set in which it occurs then $x \in \Phi(G)$.

Conversely, take any element y in $\Phi(G)$, and suppose that $G = \mathrm{gp}\{C, y\}$ where C is some complex; thus $\{C, y\}$ is a set of generators for G. To derive a contradiction, let us suppose that if $S = \mathrm{gp}\{C\}$ then $S < G$. Clearly we have $y \notin S$. We now consider the set $\mathscr{S}$ of subgroups of G:

$$\mathscr{S} = \{H : H \leqslant G,\ S \leqslant H,\ y \notin H\}.$$

Since G is finite $\mathscr{S}$ has maximal elements; in other words there is at least one subgroup H_0 in $\mathscr{S}$ not contained properly in any element of $\mathscr{S}$. We prove that such a subgroup H_0 is maximal in G. Suppose that $H_0 < H_1 \leqslant G$; by choice of H_0 we have $y \in H_1$, and so $H_1 \geqslant \mathrm{gp}\{H_0, y\} = G$. Therefore H_0 is maximal in G, and $y \notin H_0$. This contradicts the fact that y lies in $\Phi(G)$ and so in all maximal subgroups of G. Therefore $G = \mathrm{gp}\{C\}$, and y may be omitted from any generating set $\{C, y\}$ for G. □

THEOREM 9.24. *If G is a finite group then $\Phi(G)$ is nilpotent.*

Proof. We see that $\Phi(G)$ is normal, even characteristic, in G. Apply Corollary 7.28 with $K = \Phi(G)$; thus if K_p is a Sylow p-subgroup of K we have $G = KN(K_p)$. But the previous theorem shows that if $G = \mathrm{gp}\{K, N(K_p)\}$ then $G = \mathrm{gp}\{N(K_p)\}$. That is, $K_p \trianglelefteq G$. It certainly follows that each Sylow p-subgroup of $\Phi(G)$ is normal in $\Phi(G)$. By Theorem 9.08, $\Phi(G)$ is nilpotent. □

THEOREM 9.25. *The finite group G is nilpotent if and only if $\Phi(G) \geqslant \delta(G)$.*

Proof. Suppose G is finite and nilpotent. Maximal subgroups of G exist (unless G is trivial); if M is one of them then $M \trianglelefteq G$ by Theorem 9.08. In fact it is clear that G/M is of prime order since it has no proper subgroup. Thus $\delta(G) \leqslant M$ by Theorem 4.21, and it follows that $\delta(G) \leqslant \Phi(G)$.

Suppose next that $\Phi(G) \geqslant \delta(G)$; we have to prove that G is nilpotent. Let G_p be a Sylow p-subgroup of G. If $N(G_p) \neq G$ then $N(G_p) \leqslant M$ where M is a maximal subgroup of G. Thus $M \geqslant \Phi(G)$. The hypothesis that $\Phi(G) \geqslant \delta(G)$ shows that $M \geqslant \delta(G)$; and it follows that $M \trianglelefteq G$. But Corollary 7.29 shows that any subgroup containing $N(G_p)$ is self-normalizing, so M cannot be normal in G. This contradiction forces us to accept that $N(G_p) = G$, and so by Theorem 9.08 G is nilpotent. □

We return to finite p-groups.

THEOREM 9.26. *If G is a finite p-group then*

$$\Phi(G) = \mathrm{gp}\{[x,y],\, z^p : x,y,z \in G\}.$$

Proof. Put $N = \mathrm{gp}\{[x,y],\, z^p : x,y,z \in G\}$. From the proof of the previous theorem it is easy to see that N lies in every maximal subgroup of G, and so $N \leqslant \Phi(G)$.

We have yet to show that if $g \in \Phi(G)$ then $g \in N$. Suppose this is not the case for some particular g. Thus $g \in \Phi(G)$ and $g \notin N$. Then $Ng \neq N$. The group G/N is clearly elementary abelian and if its order is p^r then it is the direct product of r groups of order p. We may choose $r-1$ elements generating a subgroup of order p^{r-1} not containing Ng (this follows from our constructive proof of Theorem 5.05 describing finite abelian groups). If these elements are $a_1,\ldots, a_{r-1}$ then $\mathrm{gp}\{N, a_1,\ldots, a_{r-1}\}$ is a subgroup of G having index p and not containing g. Therefore $g \notin \Phi(G)$. It follows that $\Phi(G) = N$, as required. □

We observe that $G/\Phi(G)$ is elementary abelian, that $G/\Phi(G)$ is trivial if and only if G is trivial, and that $\Phi(G) = 1$ if and only if G is elementary abelian; these remarks apply in the case when G is a finite p-group.

THEOREM 9.27. *If G is a finite p-group with $G/\Phi(G)$ of order p^r then any generating set for G contains a generating set of r elements.*

Proof. Let $G = \mathrm{gp}\{g_1,\ldots,g_s\}$. Then

$$G/\Phi(G) = \mathrm{gp}\{\Phi(G)g_1,\ldots,\Phi(G)g_s\}.$$

Because $G/\Phi(G)$ is the direct product of r groups of order p, we can select $h_1,\ldots, h_r$ from $\{g_1,\ldots,g_s\}$ so that

$$G/\Phi(G) = \mathrm{gp}\{\Phi(G)h_1,\ldots,\Phi(G)h_r\}.$$

Therefore $G = \mathrm{gp}\{\Phi(G), h_1,\ldots,h_r\}$. But the elements of $\Phi(G)$ may be omitted from any generating set of G. We find that $G = \mathrm{gp}\{h_1, \ldots, h_r\}$, as desired. □

COROLLARY 9.28. *Let C be any complex of the finite p-group G such that C, but no proper subset of C, generates G. Then the number of elements in C is an invariant of G.* □

We repeat our caution that results such as the preceding two are peculiar to finite p-groups. Perhaps we could support it by mentioning that Z_p^∞ is a nilpotent group with no maximal subgroup, so that $Z_p^\infty = \Phi(Z_p^\infty)$, and by asserting that if $G = \mathrm{gp}\{g : g^6 = 1\}$ then the generating sets $\{g\}$ and $\{g^2, g^3\}$ for G contain different numbers of elements.

We turn now to a quite different topic, connected with the fact that a nilpotent group generated by finitely many elements each of finite order must itself be finite. The so-called Burnside problem seeks information on a finitely generated group in which each element has finite order, and in the case of p-groups we might ask whether a finitely generated p-group must be finite and so nilpotent. It is beyond the scope of this work to describe the counter-examples that provide a negative answer. Very special conditions, however, do give positive results.

DEFINITION. A group G has *exponent* n if $x^n = 1$ for each element x of G.

Even a condition on the exponent of a p-group does not give good general results. Groups of exponent 2 are trivially easy, as we have seen (in remarks after Example 4.20), and groups of exponents 3 and 4 will now be treated in an elementary fashion.

LEMMA 9.29. *If x, y, z are arbitrary elements in any group then*

(v) $[x, z, y^x][y, x, z^y][z, y, x^z] = 1.$

Proof. We simply expand each factor in (v):

$$\begin{aligned}[x, z, y^x] &= [x, z]^{-1}y^{-x}[x, z]y^x \\ &= (z^{-1}x^{-1}zx)(x^{-1}y^{-1}x)(x^{-1}z^{-1}xz)(x^{-1}yx) \\ &= (zyz^{-1}xz)^{-1}(xzx^{-1}yx).\end{aligned}$$

It should now be clear that if the two other expansions are similarly carried out then the left-hand side of (v) will cancel completely. □

LEMMA 9.30. *If a, b are elements of a group for which $a^3 = (ab)^3 = (ab^{-1})^3 = 1$ then $[a, b, b] = 1$.*

Proof. Since $(ab)^3 = 1$ we have $bab = a^{-1}b^{-1}a^{-1}$. Substitution of this in $a^{-1}bab$ gives

$$a^{-1}bab = a^{-2}b^{-1}a^{-1}.$$

But $a^3 = 1$. We therefore have

$$a^{-2}b^{-1}a^{-1} = ab^{-1}a^2.$$

Finally, $(ab^{-1})^3 = 1$ gives $ab^{-1}a = ba^{-1}b$. Therefore

$$ab^{-1}a^2 = ba^{-1}ba.$$

When combined these three equations give $a^{-1}bab = ba^{-1}ba$. Thus b commutes with $a^{-1}ba$ and so with $[b, a] = b^{-1}a^{-1}ba$ and with $[b, a]^{-1} = [a, b]$. We conclude that $[a, b, b] = 1$. □

COROLLARY 9.31. *If x, y are arbitrary elements in a group of exponent 3 then $[x, y, y] = 1$.* □

We now consider the class of groups for which $[x, y, y] = 1$ for every pair x, y of elements. We note that this condition means precisely that each element y commutes with every conjugate y^x of y and y^{-x} of y^{-1}; or alternatively that the normal closure of $\{y\}$ is abelian.

THEOREM 9.32. *If the group G satisfies the condition $[x, y, y] = 1$ for every pair of its elements x, y then G is nilpotent of class 3; $\delta(G)$ is abelian; and $\gamma_3(G)$ has exponent 3.*

Proof. If x, y, z are arbitrary elements of G then $[x, yz, yz] = 1$. We expand the commutator by means of the relations (ii) and (i) above (p. 180); thus by (ii)

$$[x, yz, yz] = [x, yz, z][x, yz, y]^z. \tag{1}$$

Application of (ii) and (i) gives

$$\begin{aligned}[x, yz, z] &= [[x, z][x, y]^z, z] \\ &= [[x, y]^z, z]\end{aligned}$$

because $[x, z, z] = 1$. Since z commutes with the conjugates of z and z^{-1} we have

$$[[x, y]^z, z] = [x, y, z]^z = [x, y, z].$$

Hence

$$[x, yz, z] = [x, y, z]. \tag{2}$$

Similar treatment is given to $[x, yz, y]$:

$$\begin{aligned}[x, yz, y] &= [[x, z][x, y]^z, y] \\ &= [x, z, y]^{[x,y]^z}[[x, y]^z, y].\end{aligned}$$

But $[x, y]^z$ can be expressed in terms of two conjugates of y, namely as $(y^{xz})^{-1}y^z$; so $[[x, y]^z, y] = 1$. Similarly $[x, z, y]$ and $[x, y]^z$ are products of conjugates of y and y^{-1}, and so commute. Thus

$$[x, yz, y] = [x, z, y]. \tag{3}$$

Combination of (1), (2), (3) gives

$$[x, y, z][x, z, y] = 1. \tag{4}$$

But by (iii) we have

$$\begin{aligned}[x, y, z] &= [[y, x]^{-1}, z] \\ &= [y, x, z]^{-[y,x]^{-1}},\end{aligned}$$

and the argument about y commuting with all its conjugates and their inverses gives

$$[x, y, z] = [y, x, z]^{-1} \tag{5}$$

at this point. We deduce from (4) and (5) that

$$[x, z, y] = [y, x, z], \tag{6}$$

and a renaming of x, y, z, which are arbitrary elements, gives

$$[y, x, z] = [z, y, x]. \tag{6'}$$

Next we apply (ii) to the terms of (v). Thus

$$\begin{aligned}[x, z, y^x] &= [x, z, y[y, x]] \\ &= [x, z, y]\end{aligned}$$

since $[x, z]$ commutes with $[y, x]$ as the usual argument shows. Hence and similarly (v) becomes

$$[x, z, y][y, x, z][z, y, x] = 1.$$

Now we use (6) and (6′):

$$[x, z, y]^3 = 1. \tag{7}$$

Take a further arbitrary element w. Then

$$\begin{aligned}[x, y, z, w] &= [x, y, w, z]^{-1} && \text{by (4)} \\ &= [w, x, y, z]^{-1} && \text{by (6)} \\ &= [y, z, [w, x]]^{-1} && \text{by (6)} \\ &= [[w, x], [y, z]] && \\ &= [y, z, w, x] && \text{by (6)} \\ &= [y, z, x, w]^{-1} && \text{by (4)} \\ &= [x, y, z, w]^{-1} && \text{by (6).}\end{aligned}$$

We conclude that $[x, y, z, w]$ has order dividing 2. But this commutator has order dividing 3 by (7). Hence

$$[x, y, z, w] = 1. \tag{8}$$

The previous calculation therefore gives, further,

$$[[y, z], [w, x]] = 1. \tag{9}$$

Now (9) proves that $\delta(G)$ is abelian. Since commutators of the form $[x, y]$ generate $\gamma_2(G)$ an easy application of (i) shows that commutators of the form $[x, y, z]$ generate $\gamma_3(G)$. By (8), $\gamma_3(G)$ is a subgroup of $\zeta(G)$, and so G has class 3. It is now clear from (7) that $\gamma_3(G)$ has exponent 3. □

COROLLARY 9.33. *A group G of exponent* 3 *is nilpotent of class* 3, *and $\delta(G)$ is abelian.* □

We observe that G need not have class 2, as Problem 14 of Chapter 11 indicates.

Our results on groups of exponent 4 are less precise.

THEOREM 9.34. *A finitely generated group of exponent 4 is finite.*

Proof. Let $G = \text{gp}\{a_1,..., a_n\}$ be such that $x^4 = 1$ for all x in G. If $G_i = \text{gp}\{a_1,..., a_i\}$ for $1 \leqslant i \leqslant n$ and $H_i = \text{gp}\{G_i, a_{i+1}^2\}$ for $1 \leqslant i < n$, then $G_{i+1} = \text{gp}\{H_i, a_{i+1}\}$ and $a_{i+1}^2 \in H_i$. These remarks show that an inductive proof of the theorem may easily be constructed once we have this key fact: if A is a group of exponent 4 with an element a and a finite subgroup B such that $A = \text{gp}\{a, B\}$ and $a^2 \in B$, then A is finite. We proceed to prove this.

A typical element g of A may be written in the form

$$b_0 ab_1 a \ldots ab_{r+1}$$

where $b_i \in B$ for $0 \leqslant i \leqslant r+1$. It suffices to prove that, for any element of A, such a form may be given with r bounded from above, for then A will have only finitely many elements. We therefore seek such a bound on r, and we suppose that the expression chosen for g has r minimal. Thus $b'_0 ab'_1 a \ldots ab'_{s+1} \neq g$ for $s < r$, whatever elements b'_i are chosen in B, and this implies that $b_i \neq 1$ for $1 \leqslant i \leqslant r$ as $a^2 \in B$.

Let i be fixed. We have $(ab_i)^4 = 1$, which implies

$$ab_i a = b_i^{-1} a^{-1} b_i^{-1} a^{-1} b_i^{-1} = c_i ac_i ab_i^{-1}$$

where $c_i = b_i^{-1} a^{-2} \in B$. Thus substitution for $ab_i a$ $(1 \leqslant i \leqslant r)$ may be made in g, and there results an expression $b'_0 ab'_1 a \ldots ab'_{r+1}$ for g in which $b'_{i+1} = b_i^{-1} b_{i+1}$. Note that $b'_{i+1} \neq b_{i+1}$ because $b_i \neq 1$. By carrying out such substitutions repeatedly, a number of expressions for g will be found. We shall describe inductively a method for obtaining i distinct expressions for g, differing one from another in the value assigned to b_i, for $1 \leqslant i \leqslant r$.

When $i = 1$ there is nothing to prove (and we may ignore the trivial case when $r = 0$). Let $1 \leqslant i < r$ and suppose that we have i expressions for g corresponding to i distinct values of b_i, while $b_{i+1},..., b_r$ are as given in the original expression for g.

Make the recommended substitution for $ab_i a$, in an arbitrary one of these expressions, so that the fixed element b_{i+1} is replaced by $b'_{i+1} = b_i^{-1} b_{i+1}$. The inductive hypothesis ensures that the resulting i values of b'_{i+1} are all distinct. Further, each is distinct from the fixed element b_{i+1} because $b_i \neq 1$, the integer r associated with g having the minimal property explained above. We therefore have $i+1$ expressions for g which differ from one another in the value of b_{i+1}, and which agree in the values of $b_{i+2},\ldots, b_r$. This completes the inductive step.

The result is r expressions for g, each with its particular value of b_r. But each b_r lies in the finite group B. Hence r cannot exceed the order of B, and this is the required upper bound on r. □

COROLLARY 9.35. *A finitely generated group of exponent* 4 *is nilpotent.*

Proof. By the theorem such a group is a finite 2-group, which must of course be nilpotent. □

Groups of exponent 3 could be treated by the method used for the case of exponent 4, provided we are content with the finitely generated case. In fact *all* groups of exponent 3 were nilpotent. This cannot be proved when the exponent is 4, because it is not in fact true.

Problems

1. Prove that the dihedral group (see Example 1.25) of order 2^{n+1} is nilpotent of class n, but not of class $n-1$, for each $n > 1$. Do the terms of the upper and lower central series coincide?

2. Prove that every proper subgroup of the group Z_p^∞ (see Example 1.09) is cyclic while Z_p^∞ itself is not cyclic. Deduce that Z_p^∞ has no maximal subgroup.

3. The group G_n is defined as $\text{gp}\{a, b : a^{p^{n+1}} = 1,\ b^{p^n} = 1,\ b^{-1}ab = a^{1+p}\}$ for $n \geqslant 1$ and p an odd prime. Prove, by considering the set of pairs $\{(\alpha, \beta) : \alpha, \beta$ residues modulo p^{n+1}, p^n respectively$\}$, or otherwise, that G_n has order p^{2n+1} and is nilpotent of class $n+1$ but not of class n.

*4. Prove that if G is any Sylow 2-subgroup of the symmetric group S_8 then G has no set of fewer than 3 generators and is nilpotent of class 4 (but not 3), while all proper subgroups and proper factor groups of G have class 3.

5. Prove that if G is nilpotent of class 3 then $\delta(G)$ is abelian.

6. Prove that the finite group G can be generated by n elements if and only if $G/\Phi(G)$ can be generated by n elements.

7. A group in which every finitely generated subgroup is nilpotent is called locally nilpotent. Prove the following properties of a locally nilpotent group G:

(i) there is a unique Sylow p-subgroup of G for each prime p;

(ii) if N is a minimal normal subgroup of G then $N \leqslant \zeta(G)$;

***(iii) if M is a maximal subgroup of G then M is normal in G.

8. Show that the set of all commutators of the form $[x_1,\ldots,x_n]$ in an arbitrary group G generates $\gamma_n(G)$.

9. Establish the true statements and discredit the false statements among the following:

(i) G is nilpotent if and only if $G/\zeta(G)$ is nilpotent.

(ii) G is nilpotent if and only if $\delta(G)$ is nilpotent.

(iii) If G has order p^4 (p is prime) then $\delta(G)$ is abelian.

(iv) If G is a finite p-group and $\delta(G)$ is cyclic then a suitable commutator generates $\delta(G)$.

(v) If each group in $\{G_\lambda : \lambda \in \Lambda\}$ is nilpotent of class n then $\prod\limits_{\lambda\in\Lambda}^{C} G_\lambda$ is nilpotent of class n.

10. Prove that $N \cap \zeta_1(G) > 1$ where N is a normal subgroup of the nilpotent group G and $N \neq 1$. Show that the normal abelian subgroup A of G is properly contained in no normal abelian subgroup if and only if it is properly contained in no abelian subgroup.

11. Show that if a nilpotent group G has $G/\delta(G)$ cyclic then G is abelian. Deduce that no nilpotent group can be generated by one class of conjugate elements unless it is cyclic.

12. Show that a finitely generated nilpotent group with finite centre is finite.

13. Prove that the group G has exponent 3 if and only if G is generated by elements of order 3 and satisfies the relation $[x, y, y] = 1$ for an arbitrary pair x, y of its elements.

14. The group G is such that $G/\gamma_2(G)$ is finitely generated. Prove that $\gamma_i(G)/\gamma_{i+1}(G)$ is finitely generated for all $n \geqslant 1$.

Deduce that if $G/\gamma_2(G)$ is finite then so is each $\gamma_i(G)/\gamma_{i+1}(G)$; and that the nilpotent group H is finite if and only if $H/\gamma_2(H)$ is finite.

15. Use the identity (v) of Lemma 9.29 to show that if A, B, C are normal subgroups of the group G then

$$[[A, B], C] \leqslant [[B, C], A][[C, A], B].$$

Deduce that if G has a central series with terms denoted (as in the text) by G_i for $0 \leqslant i \leqslant r$ then

$$[G_i, G_j] \leqslant G_{i+j+1}$$

where $0 \leqslant i \leqslant r$, $0 \leqslant j \leqslant r$, and $G_{r+k} = 1$ for $k \geqslant 0$.

16. Use the 'three-subgroup theorem', enunciated in Problem 15, to prove that $[\gamma_i(G), \zeta_i(G)] = 1$ for $i = 1, 2, ...$ in any group G.

**17. Let G be that subgroup of the symmetric group on the positive integers generated by all the permutations ρ_{ij}, for $i \geqslant 1$ and $j \geqslant 0$, where

$$\rho_{ij} = \prod_{k=1}^{p^{i-1}} (jp^i + k, jp^i + p^{i-1} + k, ..., (j+1)p^i - p^{i-1} + k)$$

and p is a fixed prime. We define subgroups G_i and H_i of G as follows:

$$G_i = \mathrm{gp}\{\rho_{ij} : j \geqslant 0\},$$
$$H_i = \mathrm{gp}\{G_1, ..., G_i\},$$

with $G_0 = H_0 = 1$.

Prove the following statements:

(i) $\rho_{rj}^{-1} G_i \rho_{rj} = G_i$ for $r \geqslant i \geqslant 1$ and $j \geqslant 0$;
(ii) $G_i \trianglelefteq G$ and $H_i \trianglelefteq G$ for $i \geqslant 1$;
(iii) G_i and H_i/H_{i-1} are abelian p-groups for $i \geqslant 1$;
(iv) every finitely generated subgroup of G is a finite p-group;
(v) no non-trivial element of H_i/H_{i-1} is central in G/H_{i-1} for $i \geqslant 1$;
(vi) G is not nilpotent.

***18. Prove that a group of order p^n has a central series

$$G = G_0 \geqslant G_1 \geqslant ... \geqslant G_{n-1} \geqslant G_n = 1$$

in which G_{i-1}/G_i has order p for $1 \leqslant i \leqslant n$. Deduce that if elements g_i are chosen so that $g_i \in G_{i-1}$, $g_i \notin G_i$ for $1 \leqslant i \leqslant n$ then each element of G has a unique expression $g_1^{\alpha_1} ... g_n^{\alpha_n}$ where $0 \leqslant \alpha_i < p$. Deduce again that there are integers β_{ij}, γ_{ijk} such that

$$g_i^p = g_{i+1}^{\beta_{i,i+1}} ... g_n^{\beta_{in}},$$
$$[g_j, g_i] = g_{j+1}^{\gamma_{i,j,j+1}} ... g_n^{\gamma_{ijn}} \quad (i < j),$$

and $0 \leqslant \beta_{ij} < p$, $0 \leqslant \gamma_{ijk} < p$.

Show that the multiplication table of G is completely determined by the β_{ij} with $1 \leqslant i < j \leqslant n$ and the γ_{ijk} with $1 \leqslant i < j < k \leqslant n$.

Establish that there are no more than $p^{(n^3-n)/6}$ groups of order p^n.

10

Soluble groups

It is now time to examine the remarkable interplay between normal structure and Sylow structure in the class of groups called finite soluble groups. We must repeat that all series considered in this book are *finite*.

DEFINITION. A group G is *soluble* if it has a normal series

$$G = G_0 \trianglerighteq G_1 \trianglerighteq \dots \trianglerighteq G_{r-1} \trianglerighteq G_r = 1$$

in which G_{i-1}/G_i is abelian for $1 \leqslant i \leqslant r$. A group with such a series of length r is said to be soluble of *length* (or *rank*) r.

The soluble groups of length 1 are precisely the abelian groups, while G is soluble of length 2 (or *metabelian*) if and only if $\delta(G)$ is abelian (by Theorem 4.21). Comparison of definitions shows that every nilpotent group is soluble. On the other hand, there are metabelian groups such as S_3 and A_4 that are not nilpotent of any class.

The concept of invariant series is specially suitable for treating the properties of soluble groups. It is elementary, and could have been introduced and discussed long ago, without the context of soluble groups, if it had seemed worth while.

DEFINITION. *An invariant series*

$$G = G_0 \geqslant G_1 \geqslant \dots \geqslant G_{r-1} \geqslant G_r = 1$$

of a group G is a normal series in which $G_i \trianglelefteq G$ for $1 \leqslant i \leqslant r$.

Thus every central series is an invariant series; of course S_3 and A_4 have invariant series but not central series.

THEOREM 10.01. *Any two invariant series for a given group have isomorphic refinements.*

Proof. Though we have not formally defined a refinement of an invariant series and isomorphism of two such series, the meanings of these terms should have been made evident by the discussion of normal series in Chapter 6. Indeed the proof of this theorem is very like that of Theorem 6.21 and will only be sketched. Let the group G have the invariant series

$$G = G_0 \geqslant G_1 \geqslant \ldots \geqslant G_{r-1} \geqslant G_r = 1,$$

$$G = H_0 \geqslant H_1 \geqslant \ldots \geqslant H_{s-1} \geqslant H_s = 1.$$

Between each G_{i-1} and G_i in the first series we insert terms $G_i(G_{i-1} \cap H_j)$ where $j = 1,\ldots,s-1$; between each H_{j-1} and H_j in the second we insert $H_j(G_i \cap H_{j-1})$ where $i = 1,\ldots,r-1$. The terms inserted are of course normal subgroups of G, and so the resulting series are invariant. An application of Zassenhaus's lemma (Theorem 6.18) proves the isomorphism of the two refinements. □

DEFINITION. A *chief series* of a group G is an invariant series

$$G = G_0 \geqslant G_1 \geqslant \ldots \geqslant G_{r-1} \geqslant G_r = 1$$

such that $G_{i-1} > G_i$ and if $G_{i-1} \geqslant N \geqslant G_i$ with $N \trianglelefteq G$ then $G_{i-1} = N$ or $N = G_i$, for $1 \leqslant i \leqslant r$. The groups G_{i-1}/G_i are called the *chief factors*.

Thus a chief series is an invariant series that cannot be refined in a non-trivial manner. Note that chief factors need not be simple groups; a glance at a chief series for A_4 will confirm this.

THEOREM 10.02. *In a group with a chief series every chief series is isomorphic to the given series.*

Proof. This result is analogous to the Jordan–Hölder theorem (6.23). Take two non-isomorphic chief series and construct isomorphic refinements as in the preceding theorem, omitting repeated members. Each chief series is isomorphic to its refinement (by the definition of chief series), so they are themselves isomorphic. □

COROLLARY 10.03. *Any two chief series of a finite group are isomorphic.* □

We note that the chief factors are thus group invariants. It is false that, conversely, a group is determined by its chief factors (even if they exist), because there are non-isomorphic groups of order 4, for example, with isomorphic chief series.

We return to soluble groups. If the group G has a normal series

$$G = G_0 \trianglerighteq G_1 \trianglerighteq \dots \trianglerighteq G_{r-1} \trianglerighteq G_r = 1$$

and if G_{i-1}/G_i is abelian, then $\delta(G_{i-1}) \leqslant G_i$ (by Theorem 4.21). This suggests the following inductive definition: put $\delta_0(G) = G$, and let $\delta_i(G) = \delta(\delta_{i-1}(G))$ for each $i \geqslant 1$. Thus $\delta_1(G) = \delta(G)$.

DEFINITION. If $\delta_s(G) = 1$ for some integer s then G is said to have the *derived series*

$$G = \delta_0(G) \geqslant \delta_1(G) \geqslant \dots \geqslant \delta_{s-1}(G) \geqslant \delta_s(G) = 1.$$

THEOREM 10.04. *The group G has an invariant series with abelian factors if and only if G has a normal series with abelian factors.*

Proof. One assertion of the theorem is trivial because any invariant series is also a normal series. So we concentrate on the less trivial part, assuming that G has a normal series

$$G = G_0 \trianglerighteq G_1 \trianglerighteq \dots \trianglerighteq G_{r-1} \trianglerighteq G_r = 1.$$

We have already noted that $\delta(G_{i-1}) \leqslant G_i$ for $1 \leqslant i \leqslant r$. If $\delta_{i-1}(G) \leqslant G_{i-1}$ for some particular i then

$$\delta_i(G) = \delta(\delta_{i-1}(G)) \leqslant \delta(G_{i-1}) \leqslant G_i.$$

But since $\delta_0(G) = G_0$ we have found an inductive proof that $\delta_i(G) \leqslant G_i$ for $0 \leqslant i \leqslant r$. In particular $\delta_r(G) = 1$.

The series

$$G = \delta_0(G) \geqslant \delta_1(G) \geqslant \dots \geqslant \delta_{r-1}(G) \geqslant \delta_r(G) = 1$$

is clearly normal with abelian factors. The proof will be complete as soon as we show that it is an invariant series. We again use induction on i to prove that $\delta_i(G) \trianglelefteq G$ for $0 \leqslant i \leqslant r$; in fact, it is just as easy to show that each $\delta_i(G)$ is a characteristic subgroup of G. When $i = 0$ the assertion is trivial. Suppose then that $i \geqslant 1$ and that $\delta_{i-1}(G)$ is characteristic. We know from Theorem 4.34 that, in any group H, $\delta(H)$ is a characteristic subgroup. Take $H = \delta_{i-1}(G)$ and apply Theorem 4.35. The conclusion is that $\delta_i(G) = \delta(\delta_{i-1}(G))$ is characteristic in G. It certainly follows that $\delta_i(G) \trianglelefteq G$ for $0 \leqslant i \leqslant r$. Thus the derived series is invariant, and the theorem is proved. □

Two statements established in the course of the above theorem are worth making explicitly.

COROLLARY 10.05. *If G has a normal series with abelian factors then $\delta_i(G) \leqslant G_i$ for $0 \leqslant i \leqslant r$, in the notation above.* □

COROLLARY 10.06. *The derived series is an invariant series.* □

The derived series is thus distinguished in much the same way as the lower central series, in their respective contexts. Note that while every nilpotent group is soluble, the converse is far from true.

Three definitions (in terms of normal or invariant or derived series) are now available for soluble groups. Pride of place was given to the one based on normal series, and this is historically justified because of the link with Galois theory (which we cannot touch on here). The alternative definitions make it clear that soluble groups have many normal subgroups, and together they provide many possible ways of proving the routine but none the less essential theorem which follows.

THEOREM 10.07. *Let G be a group with a subgroup S and a normal subgroup N. Then*

(i) *if G is soluble of length r then S is soluble of length r;*

(ii) *if G is soluble of length r then G/N is soluble of length r;*

(iii) *if G/N, N are soluble of lengths s, t respectively then G is soluble of length $s+t$.*

Proof. (i) We use the fact that $\delta_r(G) = 1$. Since $S \leqslant G$ we (clearly) have $\delta_i(S) \leqslant \delta_i(G)$ for $i = 0, 1, \ldots$, and it follows that, as required, $\delta_r(S) = 1$.

(ii) We have a normal series

$$G = G_0 \trianglerighteq G_1 \trianglerighteq \ldots \trianglerighteq G_{r-1} \trianglerighteq G_r = 1$$

for G in which the factors are abelian. We also have the normal series

$$G \trianglerighteq N \trianglerighteq 1.$$

Theorem 6.21 enables us to construct isomorphic refinements. The first refinement has abelian factor groups while the second contains N as a term. Truncation of the second therefore gives a series

$$G = H_0 \trianglerighteq H_1 \trianglerighteq \ldots \trianglerighteq H_{r-1} \trianglerighteq H_r = N$$

with abelian factors; note that the length has not increased from r because of the method of construction of the refinements. The existence of the series

$$G/N = H_0/N \trianglerighteq H_1/N \trianglerighteq \ldots \trianglerighteq H_{r-1}/N \trianglerighteq H_r/N = N$$

shows that G/N is soluble of length r because the third isomorphism theorem gives

$$(H_{i-1}/N)/(H_i/N) \cong H_{i-1}/H_i$$

for $1 \leqslant i \leqslant r$.

(iii) Take normal series with abelian factor groups for G/N and N:

$$G/N = G_0/N \trianglerighteq G_1/N \trianglerighteq \ldots \trianglerighteq G_{s-1}/N \trianglerighteq G_s/N = N,$$

$$N = N_0 \trianglerighteq N_1 \trianglerighteq \ldots \trianglerighteq N_{t-1} \trianglerighteq N_t = 1.$$

Here $G_1, \ldots, G_{s-1}$ are certain subgroups of G containing $G_s = N$. We have $G_i \trianglelefteq G_{i-1}$ for $1 \leqslant i \leqslant s$ since $G_i/N \trianglelefteq G_{i-1}/N$, and G_{i-1}/G_i is abelian because

$$(G_{i-1}/N)/(G_i/N) \cong G_{i-1}/G_i$$

by the third isomorphism theorem. The normal series

$$G = G_0 \trianglerighteq G_1 \trianglerighteq \ldots \trianglerighteq G_{s-1} \trianglerighteq N_0 \trianglerighteq N_1 \trianglerighteq \ldots \trianglerighteq N_{t-1} \trianglerighteq N_t = 1$$

proves that G is soluble of length $s+t$. □

COROLLARY 10.08. *The direct product of any finite set of soluble groups is soluble.*

Proof. It is sufficient to show that if $G = G_1 \times G_2$ with G_1, G_2 soluble then G is soluble, for only an elementary induction is lacking. But $G/G_1 \cong G_2$ and part (iii) of the theorem at once shows that G is soluble. □

The direct or cartesian product of an arbitrary set of soluble groups need not be soluble. The existence of soluble groups of length n but not $n-1$ for each positive integer n would suffice to prove this, but we shall not make the necessary digression.

An essential difference between soluble and nilpotent groups has appeared in part (iii) of the preceding theorem (note that S_3 is not nilpotent but has a normal subgroup N such that both S_3/N and N are nilpotent). Indeed (iii) makes it often convenient to prove theorems giving sufficient conditions for finite groups to be soluble by induction on the group order. Let us suppose that $\mathscr{P}$ is a property which may be possessed by finite groups and which is possessed by every subgroup and every factor group of G if it is possessed by G. For instance,† $\mathscr{P}$ might be (*a*) having odd order, or (*b*) having a π-number as order where $\pi = \{p, q\}$. Suppose further that we want to prove that a finite group with $\mathscr{P}$ is soluble. Then the required inductive proof can be constructed once the following fact is proved: the only finite simple groups with $\mathscr{P}$ are abelian.

We note in the interests of completeness that a finite group need not be soluble if all its proper subgroups are soluble. Take a finite non-abelian simple group of least order (such as the alternating group A_5 according to Problem 12, Chapter 7). Every proper subgroup is non-simple or abelian, and therefore soluble by an argument like that sketched in the last paragraph; but the original group is certainly insoluble.

The infinite cyclic group is soluble but lacks a composition series or (what is the same thing in this case) a chief series. Therefore it is reasonable to confine our attention to finite

† That either (*a*) or (*b*) really does imply solubility is a fact beyond the scope of this book to establish.

soluble groups when discussing composition and chief series. Since the only simple abelian groups are those of prime order, the composition factors of a finite soluble group have prime order. The next result concerns the chief factors.

THEOREM 10.09. *Any chief factor of a finite soluble group is an elementary abelian p-group, for some prime p.*

Proof. Let the finite soluble group G have the chief series

$$G = G_0 > G_1 > \dots > G_{r-1} > G_r = 1.$$

Each factor group G_{i-1}/G_i is a finite abelian group. If the Sylow p-subgroup S/G_i were a proper subgroup of G_{i-1}/G_i for some prime p, then S/G_i would be a characteristic subgroup of G_{i-1}/G_i and so a normal subgroup of G/G_i, by Theorem 4.35. Thus the given chief series could be properly refined, which is a contradiction. So G_{i-1}/G_i is a finite abelian p-group for some prime p. Next we consider the subgroup T/G_i of G_{i-1}/G_i generated by all the pth powers in G_{i-1}/G_i. Again, T/G_i is characteristic in G_{i-1}/G_i and so normal in G/G_i. The chief series property shows that $T = G_i$, for if $T = G_{i-1}$ then $G_{i-1} = G_i$ by the theory of abelian groups. Thus G_{i-1}/G_i has exponent p. This is the assertion of the theorem. □

COROLLARY 10.10. *A minimal normal subgroup of a finite soluble group is an elementary abelian p-group, for some prime p.*

Proof. By a minimal normal subgroup we mean, of course, a normal subgroup that contains properly no normal subgroup except 1. Such a normal subgroup N is a term of some chief series, which may be obtained by refining the invariant series $G \trianglerighteq N \trianglerighteq 1$, according to Theorem 10.01. Thus $N/1$ is a chief factor and Theorem 10.09 describes its structure. □

The last few results have completed our preparations for the Sylow-like theorems that are valid in finite soluble groups (and which in fact characterize them, though we shall not prove this). Our method is induction on the group order, and we shall, of course, use in an essential way the fact that there exist proper normal subgroups, with the usual trivial exceptions.

DEFINITION. A *Hall π-subgroup* of a finite group G is a π-subgroup whose index is not divisible by any prime in π. (π will denote a non-empty set of primes throughout this chapter.)

Thus a Hall π-subgroup is certainly a Sylow π-subgroup. We saw in Example 7.07, however, that the insoluble group A_5 has Sylow π-subgroups but not Hall π-subgroups, if $\pi = \{2, 5\}$.

The following theorem is of the greatest importance.

THEOREM 10.11. *Let G be a finite soluble group and π a set of primes. Then*

(i) *G contains Hall π-subgroups;*

(ii) *any two Hall π-subgroups of G are conjugate;*

(iii) *every π-subgroup of G lies in a suitable Hall π-subgroup.*

Proof. An induction on $|G|$ is started by the remark that the theorem is true, even trivial, when $|G|$ is a π-number. Suppose then that $|G| = mn$ where m is the largest π-number dividing $|G|$ and $n > 1$. In the general step of the induction we distinguish two cases.

The hypotheses imply that G has proper normal subgroups (compare Corollary 10.10). In the first case we suppose that K is such a subgroup of order $m_1 n_1$ with $n_1 < n$; m_1 and n_1 are to be such that they divide m, n respectively. The inductive hypothesis indicates that G/K has a subgroup, S/K say, of order m/m_1. Then S has order mn_1. On applying the inductive hypothesis to S, we find a subgroup M of order m. This is the required subgroup of G, and (i) is proved.

To establish (ii), let M_1 and M_2 be a pair of subgroups of order m. Since $M_i K/K \cong M_i/(K \cap M_i)$ for $i = 1, 2$ by the second isomorphism theorem (6.12), and since $|K \cap M_i|$ divides m_1, we see that $|M_i K/K| = (m/m_1)m_2$, where m_2 divides m_1. On the other hand $|M_i K/K|$ divides $|G/K| = mn/(m_1 n_1)$, by Lagrange's theorem (3.15), and so m_2 divides n/n_1. Therefore we have $m_2 = 1$, $|M_i K/K| = m/m_1$, and $M_i K/K$ is a Hall π-subgroup of G/K. It follows from the inductive hypothesis that $M_1 K/K$ and $M_2 K/K$ are conjugate in G/K. Thus, for some x in G,

$$M_1 K/K = (Kx)^{-1}(M_2 K/K)(Kx),$$

from which it follows that

$$M_1 K = x^{-1}(M_2 K)x, \quad \text{and} \quad x^{-1}M_2 x \leqslant M_1 K.$$

Now M_1 and $x^{-1}M_2 x$ are Hall π-subgroups of G and therefore of $M_1 K$, and a further appeal to the inductive hypothesis shows that they are conjugate in $M_1 K$ for

$$|M_1 K| = |M_1 K/K||K| = (m/m_1)(m_1 n_1) < mn = |G|.$$

Hence M_1 and M_2 are conjugate in G.

We turn to (iii). Let P be an arbitrary π-subgroup of G. Then PK/K is a π-subgroup of G/K. The inductive hypothesis shows that $PK/K \leqslant S/K$, the latter being a Hall π-subgroup of G/K. Therefore $P \leqslant S$. Again, induction shows that P lies in some Hall π-subgroup M of S and therefore of G.

The first case is now complete, and we have to consider the alternative, in which every proper normal subgroup K of G has order $m_1 n$ for some m_1; that is, $n_1 = n$ for all K. Now G certainly has minimal normal subgroups, and by Corollary 10.10 these are p-groups for some prime p. We conclude that $n = p^\alpha$ where p is a fixed prime and $\alpha > 0$; and that *all* the minimal normal subgroups have this order. Such a subgroup K is a Sylow p-subgroup of G because m and n are coprime. It follows (from Corollary 7.17) that K is the *unique* minimal normal subgroup of G. The Schur–Zassenhaus theorem (7.34) shows that there is a subgroup of order m, at this point. We shall, however, prove this in another way, for this will make the present proof more coherent.

By Corollary 10.10 G/K has a minimal normal subgroup L/K, say. Its order will be q^β where q is a different prime from p, and $\beta > 0$. Now L will have a subgroup Q of order q^β, by Sylow's first theorem, and clearly $L = KQ$. By Corollary 7.28 $G = LN(Q)$ where $N(Q)$ is the normalizer of Q in G. Therefore

$$G = LN(Q) = KQN(Q) = KN(Q).$$

Our aim now is to show that $K \cap N(Q) = 1$, for it will then follow that $|N(Q)| = |G/K| = m$, and $N(Q)$ will be the desired subgroup of order m.

Put $D = K \cap N(Q)$. We have (*a*) $D \trianglelefteq K$ because K is abelian (by Corollary 10.10); and (*b*) $D \trianglelefteq N(Q)$ because $K \trianglelefteq G$ implies $K \cap N(Q) \trianglelefteq N(Q)$ (by Corollary 6.02). It follows that $D \trianglelefteq KN(Q) = G$. But $D \leqslant K$, a minimal normal subgroup of G. Therefore $D = 1$ or $D = K$. If we had $D = K$ then $K \leqslant N(Q)$, and so $G = KN(Q) = N(Q)$; this would imply that $Q \trianglelefteq G$, but K was the unique minimal normal subgroup of G and $K \nleqslant Q$ because $p \neq q$. We therefore conclude that $D = 1$. As remarked above, this shows that $N(Q)$ has order m.

Next we show that all subgroups of order m are conjugate in G. If M is any such subgroup then $G = LM$ since $|LM|$ is divisible by both $|L| = p^\alpha q^\beta$ and $|M| = m$. The second isomorphism theorem shows that G/L and $M/(M \cap L)$ have equal orders. Since $|G| = mp^\alpha$ we find that $|M \cap L| = q^\beta$.

Now any subgroup of L which has order q^β is conjugate to Q, by Sylow's second theorem. Because $|N(Q)| = m$, and because $N(Q^x) = N(Q)^x$, its normalizer has order m and is conjugate to $N(Q)$. Therefore, if M is any subgroup of order m, $M \cap L$ is conjugate to Q, $M = N(M \cap L)$, and M is conjugate to $N(Q)$. It follows that all such subgroups M are conjugate.

Finally we prove that any π-subgroup S of G lies in a Hall π-subgroup. Let the order of S be m' where $m' < m$. It is easy to see that $|SK| = m'p^\alpha < |G|$. Because M and SK have coprime indices in G, it follows from Theorem 3.20 (ii) that $|M \cap SK|$ is the greatest common divisor of $|M|$ and $|SK|$, which is m'. Thus S and $M \cap SK$ are subgroups of SK with equal orders. The inductive hypothesis shows that they are conjugate in SK. Therefore S lies in a suitable conjugate of M, a Hall π-subgroup. □

This basic result on the structure of soluble groups makes possible a considerable theory, of which we give only a specimen.

Definition. A set $\mathscr{S}$ of Sylow p-subgroups of the finite group G (one for each prime p dividing $|G|$) is a *Sylow basis* if $PQ = QP$ for all $P, Q \in \mathscr{S}$.

Definition. Two Sylow bases $\mathscr{S}$, $\mathscr{T}$ of the finite group G are *conjugate* if there is an element x in G for which $P \in \mathscr{S}$ implies $P^x \in \mathscr{T}$.

DEFINITION. The *normalizer* of a Sylow basis $\mathscr{S}$ is the set of elements of G normalizing each element of $\mathscr{S}$.

THEOREM 10.12. *If G is a finite soluble group, then*

(i) *G has at least one Sylow basis;*

(ii) *any two Sylow bases are conjugate;*

(iii) *the normalizer of a Sylow basis is nilpotent.*

Proof. (i) Let $\pi = \{p_1, ..., p_r\}$ be the set of primes dividing $|G|$, and let π_i be π less p_i for $1 \leqslant i \leqslant r$. Theorem 10.11 ensures that G contains a Hall π_i-subgroup H_i for each i. We put $P_j = \bigcap_{i \neq j} H_i$. Since the indices in G of the H_i are coprime, an obvious extension of Theorem 3.20 (ii) shows that P_j is a Sylow p_j-subgroup of G.

To prove that $P_j P_k = P_k P_j$ for $j \neq k$, it is sufficient to show that $P_j P_k \leqslant G$, by Theorem 6.08. It is clear that P_j and P_k are contained in $\bigcap H_i$ if the intersection is taken over all i except j, k. Therefore $P_j P_k$ is contained in this intersection. The usual argument about orders gives equality, so that $P_j P_k$ is a subgroup of G.

(ii) Take two Sylow bases $\mathscr{S} = \{P_i\}, \mathscr{T} = \{Q_i\}$ for G. Construct Hall π_i-subgroups H_i, K_i from $\mathscr{S}$, $\mathscr{T}$ respectively, by defining $H_i = \prod_{j \neq i} P_j$ and $K_i = \prod_{j \neq i} Q_j$. Let $\mathscr{S}^x = \{P_i^x\}$ be a conjugate of $\mathscr{S}$, with the greatest possible number of H_i^x in common with $\{K_i\}$. If $\mathscr{S}^x \neq \mathscr{T}$, then some H_i^x does not coincide with the corresponding K_i. But it is clear that H_i^x and K_i are Hall π_i-subgroups of G and are therefore conjugate in G by Theorem 10.11: $H_i^{xy} = K_i$, say. Since $G = H_i^x P_i^x$, we may even assume that $y \in P_i^x$. But then $H_j^{xy} = H_j^x$ since $P_i^x \leqslant H_j^x$ if $j \neq i$. We now have $\{H_i^{xy}\}$ containing one more element in common with $\mathscr{T}$ than $\mathscr{S}^x$, contradiction to the choice of x. We therefore have $\mathscr{S}^x = \mathscr{T}$, as required.

(iii) Let N be the normalizer of the Sylow basis $\mathscr{S}$ of G. Clearly N is a subgroup of G. Let P_i be the p_i-subgroup in $\mathscr{S}$, and let $g \in N$ have p_i-power order. Since $g \in N(P_i)$ we have $g \in P_i$, by Corollary 7.19. Therefore $N \cap P_i$ is the Sylow p_i-

subgroup of N. The fact that the Sylow p_i-subgroup of N is unique shows that N is nilpotent (by Theorem 9.08 and Corollary 7.17). □

We pass on to what is clearly a special class of soluble groups.

DEFINITION. A group is *supersoluble* if it has an invariant series with cyclic factors.

Not every soluble group is supersoluble; this is established by the example A_4. Neither is every nilpotent group supersoluble; consider Z_p^∞. But finite p-groups and finite nilpotent groups are supersoluble:

THEOREM 10.13. *A finitely generated nilpotent group is supersoluble.*

Proof. If G is nilpotent then G has a (finite) central series. The factors are finitely generated because all subgroups of G are finitely generated (Theorem 9.16). These factors are therefore direct products of cyclic groups, by Theorem 5.14. Because of the central property of the series for G, it can be refined to a central series with cyclic factors without losing finiteness of length. This is an invariant series, so G is supersoluble. □

We note that supersoluble groups need not be nilpotent, for S_3 is not nilpotent. For finite groups, supersolubility occupies an interesting intermediate position between nilpotence and solubility.

Next we shall consider cyclic factors of invariant series in detail. A lemma is required.

LEMMA 10.14. *The automorphism group of a cyclic group is abelian.*

Proof. Let α, β be automorphisms of the cyclic group G with generator g. Then $g\alpha = g^u$, $g\beta = g^v$ for some integers u, v; it follows that $g\alpha\beta = g^{uv} = g\beta\alpha$. Thus $x\alpha\beta = x\beta\alpha$ for all x in G, so $\alpha\beta = \beta\alpha$. Therefore the full automorphism group of G is abelian. □

THEOREM 10.15. *Let the group G have the invariant series*

$$G \geqslant \delta(G) = K_0 > K_1 > \dots > K_{r-1} > K_r = 1$$

where K_{i-1}/K_i is cyclic for $1 \leqslant i \leqslant r$. Then

$$\delta(G) = K_0 > K_1 > \dots > K_{r-1} > K_r = 1$$

is a central series for $\delta(G)$.

Proof. We have to show that K_{i-1}/K_i lies in the centre of $\delta(G)/K_i$ for $1 \leqslant i \leqslant r$. Now K_{i-1}/K_i is cyclic, and the automorphisms arising from conjugation by elements xK_i, yK_i of G/K_i commute, by the lemma. Therefore $[x, y]K_i$ induces the trivial automorphism on K_{i-1}/K_i. Therefore $\delta(G)/K_i$ and K_{i-1}/K_i commute element by element, and this is the desired result. □

Corollary 10.16. *If the group H has $\delta_i(H)/\delta_{i+1}(H)$ cyclic but not trivial for some fixed $i \geqslant 1$, then there is no normal subgroup N of H such that $N < \delta_{i+1}(H)$ and $\delta_{i+1}(H)/N$ is cyclic.*

Proof. Suppose that there is in fact such a subgroup N, and apply the theorem with $G = \delta_{i-1}(H)/N$. We find that $\delta_{i+1}(H)/N$ is central in $\delta_i(H)/N$, since

$$\{\delta_i(H)/N\}/\{\delta_{i+1}(H)/N\} \cong \delta_i(H)/\delta_{i+1}(H)$$

(by the third isomorphism theorem), and this group is cyclic by hypothesis. By Theorem 4.27, $\delta_i(H)/N$ is abelian. By Theorem 4.21, $N \geqslant \delta_{i+1}(H)$. We therefore find that $N = \delta_{i+1}(H)$, a contradiction. There is, therefore, no such subgroup as N. □

Corollary 10.17. *If $\delta_i(H)/\delta_{i+1}(H)$ and $\delta_{i+1}(H)/\delta_{i+2}(H)$ are both cyclic for some group H and for some $i \geqslant 1$, then $\delta_{i+1}(H) = \delta_{i+2}(H)$.* □

This has a number of interesting implications; for example, $\delta(G) \cong S_3$ is impossible for any group G.

Corollary 10.18. *If G is supersoluble then $\delta(G)$ is nilpotent.*

Proof. Let

$$G = G_0 \geqslant G_1 \geqslant \dots \geqslant G_{r-1} \geqslant G_r = 1$$

with each G_{i-1}/G_i cyclic be an invariant series for G. Put $K_i = \delta(G) \cap G_{i+1}$ for $0 \leqslant i < r$; thus $K_0 = \delta(G)$ since G_0/G_1 is abelian and so $\delta(G) \leqslant G_1$ by Theorem 4.21. Now

$$\delta(G) = K_0 \geqslant K_1 \geqslant \dots \geqslant K_{r-2} \geqslant K_{r-1} = 1$$

is an invariant series for $\delta(G)$ because $G_{i+1} \trianglelefteq G$ and so $K_i \trianglelefteq \delta(G)$; and K_{i-1}/K_i is cyclic for $1 \leqslant i < r$ because

$$K_{i-1}/K_i = (\delta(G) \cap G_i)/(\delta(G) \cap G_{i+1})$$
$$\cong (\delta(G) \cap G_i)G_{i+1}/G_{i+1}$$

(by the second isomorphism theorem) and this is a subgroup of the cyclic group G_i/G_{i+1}. Omit repetitions in this invariant series for $\delta(G)$ and apply Theorem 10.15, concluding that $\delta(G)$ is nilpotent. □

Some alteration of invariant series of a supersoluble group is possible.

THEOREM 10.19. *A supersoluble group has an invariant series in which every factor is cyclic of infinite or prime order.*

Proof. Let G have the invariant series

$$G = G_0 \geqslant G_1 \geqslant \ldots \geqslant G_{r-1} \geqslant G_r = 1$$

in which G_{i-1}/G_i is cyclic for $1 \leqslant i \leqslant r$. If a factor is infinite, we disregard it; if it is trivial, we omit it. Otherwise we examine its subgroups with a view to refining the invariant series which we started with. If $|G_{i-1}/G_i| = p_1 \ldots p_t$ where $p_1, \ldots, p_t$ are (not necessarily distinct) primes, then G_{i-1}/G_i has precisely one subgroup H_{j-1}/G_i of order $p_j \ldots p_t$ for $1 \leqslant j \leqslant t$; remember that G_{i-1}/G_i is cyclic. Such a subgroup H_{j-1}/G_i must be characteristic in G_{i-1}/G_i, and therefore normal in G/G_i, by Theorem 4.35. Therefore $H_{j-1} \trianglelefteq G$ and H_{j-1}/H_j has prime order for $1 \leqslant j \leqslant t$ (with $H_t = G_i$). It follows that

$$G = G_0 \geqslant \ldots \geqslant G_{i-1} \geqslant H_1 \geqslant \ldots \geqslant H_{t-1} \geqslant G_i \geqslant \ldots \geqslant G_r = 1$$

is an invariant series for G with H_{j-1}/H_j of prime order. Sufficient repetition of this process will produce an invariant series of the required kind for G. □

COROLLARY 10.20. *A chief series for a finite supersoluble group is also a composition series.*

Proof. This is now obvious from the definitions. □

LEMMA 10.21. *Let G be a group of order pq where the primes p, q are such that $p > q$. Then G has a characteristic subgroup of order p.*

Proof. We use the Sylow theorems. A subgroup P of order p has $1+kp$ conjugates for some k; as usual $1+kp$ divides pq, and since $p > q$ we see that $1+kp$ divides 1, that is $k = 0$. There is, therefore, only one subgroup of order p, and it is characteristic by Corollary 7.18. □

(Note that a similar argument applies to the Sylow q-subgroup, showing that G is abelian unless $p \equiv 1$ modulo q.)

THEOREM 10.22. *Let the group G have an invariant series*

$$G = G_0 \geqslant G_1 \geqslant \dots \geqslant G_{i-1} \geqslant G_i \geqslant G_{i+1} \geqslant \dots \geqslant G_{r-1} \geqslant G_r = 1$$

in which G_{i-1}/G_i, G_i/G_{i+1} have prime orders p, q respectively. Then G has an invariant series

$$G = G_0 \geqslant G_1 \geqslant \dots \geqslant G_{i-1} \geqslant H \geqslant G_{i+1} \geqslant \dots \geqslant G_{r-1} \geqslant G_r = 1$$

in which G_{i-1}/H, H/G_{i+1} have orders q, p respectively provided $p > q$.

Proof. The group G_{i-1}/G_{i+1} has order pq and therefore contains a characteristic subgroup H/G_{i+1} of order p, by the lemma. The second invariant series is now constructed in the obvious fashion. □

COROLLARY 10.23. *Let G be a finite supersoluble group and let $|G| = p_1^{\alpha_1} \dots p_u^{\alpha_u}$ where $p_1 > p_2 > \dots > p_u$, these p_i being prime. Then G has normal subgroups of orders $p_1^{\alpha_1} \dots p_i^{\alpha_i}$ for $1 \leqslant i \leqslant u$.*

Proof. Let

$$G = G_0 \geqslant G_1 \geqslant \dots \geqslant G_{r-1} \geqslant G_r = 1$$

be an invariant series for G. By Theorem 10.19 we may suppose that each G_{i-1}/G_i has prime order. By repeated application of Theorem 10.22 we may assume that the last α_1 factors of the series have order p_1. This proves that G has a normal (even characteristic) subgroup of order $p_1^{\alpha_1}$. An obvious induction (on u) would complete a rigorous proof of the theorem, but details of this are naturally omitted. □

Problems

1. Prove that two finite abelian groups have the same order if and only if their chief series are isomorphic.

2. Let G be a finite group. Prove that the orders of the chief factors of a subgroup divide the orders of the chief factors of G.

Prove also that the orders of the chief factors of a factor group are a subset of the orders of the chief factors of G.

3. Prove that a finite group is soluble if and only if every factor group other than 1 has a normal abelian subgroup other than 1.

4. Prove that $\delta(G_1 \times G_2) \cong \delta(G_1) \times \delta(G_2)$. Deduce that if G_1 and G_2 are soluble of length r then so is $G_1 \times G_2$.

5. Let G be a group of order p^2q^2 where p, q are primes and $p > q$. Prove that G has a normal Sylow p-subgroup if $p > 3$. Is this the case when $p = 3$? Show that G is always soluble.

*6. The group G has the form AB for certain abelian subgroups A, B. Prove (by considering commutators or otherwise) that G is metabelian.

Show also that if $G \neq 1$, $G \neq A$, $G \neq B$ then one of A, B contains a proper normal subgroup of G; and that one of A, B is contained in a proper normal subgroup of G.

7. Find all the Hall π-subgroups of A_5. Does there exist a Sylow basis?

Prove that all the subgroups of order 6 are conjugate, and investigate conjugacy in the set of subgroups of order 10.

8. Show that the symmetric group S_4 is soluble, and find all its Sylow bases.

9. Prove that the subgroup G of S_7 generated by $\{(1234567), (243756)\}$ has order 42 and is soluble.

Find a Sylow 2-subgroup and a Sylow 3-subgroup that do not lie in the same Sylow basis; and find two Sylow bases, each containing one of these subgroups.

10. Prove that a finite soluble group is nilpotent if and only if its Hall π-subgroups of prime-power index are all normal.

11. Let h be the number of Hall π-subgroups of a finite soluble group G. Prove that h is the product of a set of numbers with both the following properties:
   (i) each is congruent to 1 modulo some prime in π;
   (ii) each divides the order of some chief factor of G.

12. Is it true that a group with a composition series also has a chief series? (As usual, all series are finite.)

13. Prove the following statements about the supersoluble groups G and H:
   (i) Every subgroup of G is supersoluble.
   (ii) Every factor group of G is supersoluble.
   (iii) $G \times H$ is supersoluble.
   (iv) G is finitely generated.
   (v) If $G \cong K/N$, where N is a cyclic normal subgroup of K, then K is supersoluble.

14. The group G has order $2p^n$ where p is an odd prime. Prove that G is soluble, and that
   (i) a minimal normal subgroup M of p-power order lies in the centre of the Sylow p-subgroup P;
   (ii) if $M = P$ then M is cyclic;
   (iii) G is supersoluble.

*15. The group G is the subgroup of the symmetric group S_8 generated by $\{(12)(35)(47)(68), (2345687), (346)(578)\}$. Prove that the normalizers of the Sylow systems have order 3 and are contained in cyclic subgroups of order 6. Is G supersoluble?

16. Prove the true and disprove the false statements among the following.
   (i) $\text{gp}\{a, b : a^{-1}ba = b^2\}$ is soluble.
   (ii) If $G = AB$ where A, B are soluble subgroups then G is soluble.
   (iii) If all the Sylow subgroups of the finite group G are abelian then G is soluble.
   (iv) A soluble group is metabelian if and only if it has a chief series.
   (v) If all proper subgroups of G are supersoluble then G is supersoluble.

17. The soluble group G has a normal series in which every factor is finitely generated abelian. Prove that G has a normal series in which every factor is either the direct product of a finite number of cyclic groups of prime order or a similar product of cyclic groups of infinite order. Show that the number of infinite factors in such a series is a group invariant.

*18. The finite group G has a normal Hall π-subgroup K, and so by the Schur–Zassenhaus theorem G has a subgroup of order $|G/K|$. Prove that if either K or G/K is soluble then any two such subgroups are conjugate.

11

Survey of examples

IN this final chapter we return to the subject of Chapter 1: examples of groups. We shall not only discuss our earlier examples in the light of the theory we have developed, but also produce further examples that have been promised or that are relevant to various facts in this book. We should perhaps warn the reader that the examples of Chapter 1 are well-known groups and have reasonable properties; further experience may perhaps lead him to the conclusion that most groups tend to have unexpected and even discouraging aspects, and indeed some intransigent counter-examples have already appeared among the problems.

Perhaps our first duty is to find all groups of small order. There are no surprises, for we have become familiar with most of them in one context or another.

Groups of order p or p^2 are abelian (see Corollary 4.30), and abelian groups are treated as in Theorem 5.14. There is one isomorphism class of order p, and two of order p^2. The possible types in the latter case are $\{2\}$ and $\{1, 1\}$; that is to say, the group is cyclic or the direct product of two groups of order p. These remarks are all that is necessary for the groups of order less than 16, except those of orders 6, 8, 10, 12, 14, 15.

We proceed to groups of order $2p$, where p is an odd prime; this will deal with 6, 10, and 14 as one case. As for abelian groups of order $2p$, Theorem 5.01 shows that such a group is the direct product of a group of order 2 and a group of order p, and this is of course cyclic. Suppose then that G is a non-abelian group of order $2p$. By Sylow's first theorem G contains elements a, b

of orders 2, p respectively. By Sylow's other theorems $\text{gp}\{b\}$ has only one conjugate in G. Therefore $b^a = b^\alpha$, and since $a^2 = 1$ we have $b^{\alpha^2} = b$. It follows that $\alpha \equiv -1$ modulo p. We have the following relations in G:

$$a^2 = 1, \quad b^p = 1, \quad b^a = b^{-1}.$$

Conversely, these relations define a group of order $2p$, namely the dihedral group of Example 1.25. There are, therefore, two isomorphism classes of groups of order $2p$.

Theory shows that there are three distinct abelian groups of order 8; the types are $\{3\}, \{2, 1\}, \{1, 1, 1\}$. A non-abelian group G of order 8 must contain an element of order 4, for otherwise every element would have order 2 or 1 which implies that G is abelian (see remarks following Example 4.20). This element b of order 4 generates a subgroup of index 2 in G, which is normal (see Example 4.01). If $a \notin \text{gp}\{b\}$ then $b^a = b^\alpha$. Since b has order 4, $\alpha = \pm 1$. Further, $a^2 \in \text{gp}\{b\}$, and of course a has order 4 or 2 (if a has order 8 then G is cyclic); so $a^2 = 1$ or $a^2 = b^2$. Therefore two possible sets of relations emerge:

$$b^4 = 1, \quad a^2 = 1, \quad b^a = b^{-1};$$

$$b^4 = 1, \quad a^2 = b^2, \quad b^a = b^{-1}.$$

These correspond to non-isomorphic groups, namely the dihedral and quaternion groups of order 8 (see Examples 1.25 and 1.30). Non-isomorphism follows from the fact that the former has three elements of order 2 and the latter only one.

Next we discuss groups of order 12. Here each Sylow 3-subgroup has order 3, while each Sylow 2-subgroup may be either of the two groups of order 4. In the abelian case, therefore, there are just two possibilities: $Z_3 \times Z_4$ or $Z_3 \times Z_2 \times Z_2$, where Z_n denotes the cyclic group of order n. Note that the former of these groups is simply Z_{12}. In the non-abelian case, Example 7.10 shows that a group G of order 12 still has one normal Sylow subgroup. If this is the Sylow 2-subgroup the reader will be able to verify that $G = A_4$ is the only solution to our problem, because the cyclic group of order 4 has no automorphism of order 3. Suppose finally that the Sylow 3-subgroup of G is normal. There are then two possibilities for G, depending on the structure of its Sylow 2-subgroup, and the following presentations for them can be found:

$$\mathrm{gp}\{a, b : a^3 = 1,\ b^4 = 1,\ a^b = a^{-1}\},$$

$$\mathrm{gp}\{a, b, c : a^3 = 1,\ b^2 = c^2 = [b, c] = 1,\ a^b = a^{-1},\ a^c = a\}.$$

We leave the details to the reader, noting that the latter group is just $S_3 \times Z_2$.

Finally we glance at the groups of order 15. Lemma 10.21 and a remark following show that such a group must be abelian, and we soon conclude that it is in fact cyclic.

We now reconsider the examples of Chapter 1, starting with abelian groups. Examples 1.01, 1.02, 1.03, 1.09, 1.10, which were Z, Q, R, Z_p^∞, Z_n respectively, are of fundamental importance. Their structure can hardly be given in simpler terms, rather they are themselves used to elucidate the structure of more complicated groups.

Example 1.04 was the additive group C of complex numbers. This is not of such great significance, because it is now obvious that it is isomorphic to $R \oplus R$ (in additive notation).

The multiplicative group Q^* of positive rationals, in Example 1.05, can also be decomposed. It is generated by the set $\{p_1, p_2, \ldots, p_n, \ldots\}$ where p_n is the nth prime, because every positive rational (except perhaps 1) can be written in the form $p_{i_1}^{\alpha_1} p_{i_2}^{\alpha_2} \ldots p_{i_r}^{\alpha_r}$ where the α_i are non-zero integers. In fact Theorem 6.30 shows that Q^* is the direct product of the subgroups generated by $p_1, p_2, \ldots, p_n, \ldots$ respectively. The group of non-zero rationals is the direct product of Q^* and the group generated by -1 (which has order 2).

The multiplicative group of positive reals R^* (Example 1.06) is isomorphic to the additive group R; this follows from well-known properties of the logarithmic function, for we may define an isomorphism ϕ from R^* onto R by $x\phi = \ln x$ (logarithm to any positive base except 1). The multiplicative group of non-zero reals is the direct product of R^* and the group of order 2.

Next we examine C^*, the multiplicative group of non-zero

complex numbers, in Example 1.07. Of course, such a number z can be expressed in the form $re^{2\pi\theta i}$ where $r > 0$ and $0 \leqslant \theta < 1$; some thought along with the use of de Moivre's theorem shows that C^* is isomorphic to $R^* \times R/Z$, and as we have seen this is isomorphic to $R \times R/Z$. We shall not discuss R/Z further; we merely note that this is the abstract group of Example 1.08.

We turn to a brief discussion of the various groups of congruence mappings. It will be seen that Example 1.26 is the algebraic description of Example 1.16, the group of translations of $\mathscr{L}$; this group is isomorphic in an obvious manner to R. Similarly Example 1.27 arises from Example 1.17. Now it is easy to verify that $\phi_{-a}\psi_0\phi_a = \psi_a$, and $\psi_0^2 = 1$; so Example 1.17 is generated by R and the further element ψ_0 of order 2. Note that $\psi_0^{-1}\phi_a\psi_0 = \phi_{-a}$; that is, ψ_0 induces the automorphism of R according to which every element is inverted.

It should now be clear that the groups of Examples 1.18 and 1.21 are isomorphic to $R \times R$ and to $R \times R \times R$ respectively. Example 1.19 is another representation of R/Z, since each rotation of $\mathscr{P}$ about a fixed point can be specified by some θ in $0 \leqslant \theta < 2\pi$, and addition is modulo 2π. The group (given in Example 1.22) of all rotations of $\mathscr{S}$ about a fixed line is isomorphic to this. These groups may also be presented as groups of matrices. We leave this task, and further investigation of Examples 1.20, 1.23, and 1.24, to the reader.

Example 1.25 is a finite group to which frequent reference has been made, and we now write down generators and relations for it. Let b denote any rotation through $2\pi/n$ and a any reflection; then it will be found that

$$a^2 = 1, \quad b^n = 1, \quad a^{-1}ba = b^{-1}.$$

Conversely, the group defined abstractly by these generators and relations coincides with the dihedral group of order $2n$. It will be found that if n is odd a has n conjugates and these are all reflections; and that the powers of b are all rotations. The most important special cases arise where n is a prime or is a power of 2.

The quaternion group (Example 1.30) is another example of importance. In Chapter 8 we saw that it may be presented abstractly as
$$\text{gp}\{a, b : a^2 = b^2 = (ab)^2\}.$$

Example 1.31 is of course isomorphic to the additive group R of reals, while Example 1.32 is isomorphic to the direct sum of mn copies of R, provided the elements of the matrices are real numbers.

Example 1.11 will now be generalized and discussed at some length. Let m be an integer greater than 1. The set $\{[n]\}$ of residue classes modulo m, where $1 \leqslant n < m$ and n is coprime to m, is the object of interest; we denote it by G_m. A routine proof, using the Euclidean algorithm, shows that G_m is an abelian group with identity element $[1]$ under the usual multiplication of residue classes—for details compare Example 1.11. It is now reasonable to ask for the structure of G_m in terms of the theory of finite abelian groups given in Chapter 5. For the present we use multiplicative notation for all groups.

At this point we indicate another way of viewing G_m.

THEOREM 11.01. *The automorphism group $A(Z_m)$ of the cyclic group Z_m of order m is isomorphic to G_m.*

Proof. Let z be a generator of Z_m. It follows from the equation $(z^k)\alpha = (z\alpha)^k$ that any automorphism α of Z_m is completely specified by $z\alpha$. Another easy fact is that $z\alpha_n = z^n$ determines an automorphism α_n of Z_m if n is non-zero and prime to m; and all the automorphisms of Z_m are determined by such values of n in $1 \leqslant n < m$. Consider the correspondence in which α_n is paired with $[n]$. That this is an isomorphism of $A(Z_m)$ with G_m is evident. □

Note that Lemma 10.14 has now been superseded.

LEMMA 11.02. *Let $H = H_{p_1} \times \ldots \times H_{p_n}$ where $p_1, \ldots, p_n$ are distinct primes and H_p denotes the Sylow p-subgroup of H. Then*
$$A(H) \cong A(H_{p_1}) \times \ldots \times A(H_{p_n}).$$

Proof. We note that each H_p is a characteristic subgroup of H. If therefore an automorphism α of H is given, then α determines by restriction an automorphism α_i of H_{p_i}, for each i. Conversely, suppose that an automorphism α_i of H_{p_i} is given, for each i. It is then easy to see that an automorphism α of H is determined by
$$(h_1,\ldots, h_n)\alpha = (h_1\alpha_1,\ldots, h_n\alpha_n),$$
where $h_i \in H_{p_i}$; one merely verifies that α is well defined, is a homomorphism, is one–one, and is onto H.

We now turn to the direct product criterion of Theorem 4.39. It is clear that the $A(H_{p_i})$ generate $A(H)$, from the remarks above. The requirements (i) and (iii) are evidently satisfied, and therefore $A(H) \cong A(H_{p_1})\times\ldots\times A(H_{p_n})$. □

Corollary 11.03. *If $m = p_1^{\beta_1}\ldots p_r^{\beta_r}$ where $p_1,\ldots, p_r$ are distinct primes, then $A(Z_m) \cong A(Z_{p_1^{\beta_1}})\times\ldots\times A(Z_{p_r^{\beta_r}})$.*

Proof. Use Theorem 5.01 and Lemma 11.02. □

Corollary 11.04. *If $m = p_1^{\beta_1}\ldots p_r^{\beta_r}$ where $p_1,\ldots, p_r$ are distinct primes, then $G_m \cong G_{p_1^{\beta_1}}\times\ldots\times G_{p_r^{\beta_r}}$.*

Proof. Use Theorem 11.01 and Corollary 11.03. □

Theorem 11.05. *If p is an odd prime then G_{p^β} is cyclic of order $(p-1)p^{\beta-1}$. If $p = 2$ then G_{2^β} is the direct product of a cyclic group of order $2^{\beta-2}$ and a cyclic group of order 2 unless $\beta = 1$ and G_2 is trivial.*

Proof. It is easy to see that $p^{\beta-1}$ of the numbers $\{n: 1 \leqslant n \leqslant p^\beta\}$ are divisible by p, so $|G_{p^\beta}| = (p-1)p^{\beta-1}$, whether p is even or odd.

Take p odd, to start with. It will clearly suffice to produce elements of G_{p^β} with orders $p-1$ and $p^{\beta-1}$ respectively, for their product will be a generator of G_{p^β}.

Consider $[1+p]$. We have, by the binomial theorem,
$$(1+p)^{p^{\beta-1}} \equiv 1 \text{ modulo } p^\beta,$$
$$(1+p)^{p^{\beta-2}} \equiv 1+p^{\beta-1} \text{ modulo } p^\beta$$
if $\beta > 1$ and $p > 2$; so in this case $[1+p]$ has order $p^{\beta-1}$.

To produce an element of order $p-1$ is rather more difficult. Let d be a factor of $p-1$ and consider the congruence

$$x^d-1 \equiv 0 \text{ modulo } p.$$

We shall need the well-known fact† that this has at most d solutions modulo p. There are, therefore, at most d elements of order dividing d in the multiplicative group of residues modulo p, for each divisor d of $p-1$. Application of the fundamental theorem (5.07) on finite abelian groups, therefore, shows that G_p is cyclic. If $[a]$ generates G_p then $a^{p-1} \equiv 1$ modulo p, and so the element $[a^{p-1}]$ of G_{p^β} has p-power order, by binomial calculations similar to those above. A suitable power of $[a]$ will have order $p-1$ as element of G_{p^β}, and this is what we want.

Next we work with $p = 2$, neglecting the trivial case $\beta = 1$. Indeed we take $\beta > 2$, for it is easy to see that $A(Z_4)$ has order 2. A binomial calculation shows that $[5]$ has order $2^{\beta-2}$ as element of G_{2^β}:

$$5^{2^{\beta-2}} = (1+4)^{2^{\beta-2}} \equiv 1 \text{ modulo } 2^\beta,$$

$$5^{2^{\beta-3}} = (1+4)^{2^{\beta-3}} \equiv 1+2^{\beta-1} \text{ modulo } 2^\beta.$$

Clearly $[-1]$ has order 2. Further, $[-1]$ is not a power of $[5]$; if it were its order would force us to

$$[-1] = [5]^{2^{\beta-3}},$$

$$-1 \equiv 1+2^{\beta-1} \text{ modulo } 2^\beta,$$

which is impossible as $\beta > 2$. It is now evident that G_{2^β} is the direct product of gp{[5]} and gp{[−1]}. □

COROLLARY 11.06. *Each of the isomorphic groups G_m and $A(Z_m)$, where $m = p_1^{\beta_1} \ldots p_r^{\beta_r}$, $p_1 < \ldots < p_r$ are primes, and $\beta_1, \ldots, \beta_r$ are positive, is the direct product of abelian groups of orders $(p_1-1)p_1^{\beta_1-1}, \ldots, (p_r-1)p_r^{\beta_r-1}$ respectively. These are cyclic when p_i is odd, and have type $\{\beta_1-2, 1\}$ when $p_1 = 2$.*

Proof. Use Theorem 11.01, Corollary 11.04, and Theorem 11.05. □

† See G. Birkhoff and S. MacLane, *A survey of modern algebra* (third edition, Macmillan, 1965), p. 58.

Example 11.07. G_m need not be cyclic; for instance G_{15} is the direct product of cyclic groups of orders 2 and 4.

We pass on to an examination of Examples 1.13, 1.14, and 1.15, which were permutation groups. We note that any group G is a subgroup of a suitable symmetric group S_X, for Theorem 2.18 shows that $G \leqslant S_X$ where X is the set of elements in G. In particular, if G has finite order n then $G \leqslant S_n$.

We now generalize Example 6.22, according to which A_5 is simple.

THEOREM 11.08. *The alternating group A_n is simple if $n \geqslant 5$.*

Proof. We proceed by induction on n, starting off with the known fact that A_5 is simple. Let $n \geqslant 6$ and suppose that A_{n-1} is simple. Let N be a normal subgroup of A_n other than 1; we want to prove that $N = A_n$.

We first show that N contains a non-trivial permutation which leaves fixed an element of the set $\{1,..., n\}$ on which A_n acts. Suppose not. Then $\phi \in N$ where

$$a\phi = b, \quad c\phi = d$$

and a, b, c, d are distinct elements of $\{1,..., n\}$. Since $n \geqslant 6$, A_n contains ψ such that

$$a\psi = b, \quad b\psi = a, \quad c\psi = d, \quad d\psi = e$$

where e is different from a, b, c, d. Now $\psi^{-1}\phi\psi \in N$, and

$$b(\psi^{-1}\phi\psi) = a, \quad d(\psi^{-1}\phi\psi) = e.$$

We also have $\phi\psi^{-1}\phi\psi \in N$, and

$$a(\phi\psi^{-1}\phi\psi) = a, \qquad c(\phi\psi^{-1}\phi\psi) = e.$$

Thus $\phi\psi^{-1}\phi\psi$ is a non-trivial element of N leaving the symbol a fixed. This is a contradiction. We conclude that N contains a non-trivial element fixing some symbol in $\{1,..., n\}$.

Suppose (without loss of generality) that $\phi \in N$ and $1\phi = 1$. It is easy to see that the set of all elements in A_n which fix 1 forms a subgroup $S(1)$ which is isomorphic to A_{n-1}. Thus we have $N \cap S(1) \neq 1$, and, of course, $N \cap S(1)$ is a normal subgroup of $S(1)$. But $S(1)$ is simple because A_{n-1} is simple by the inductive hypothesis. Therefore $N \cap S(1) = S(1)$; that is, $N \geqslant S(1)$.

Next $N \geqslant S(i)$ for each i in $\{1,..., n\}$; for if $i\psi = 1$ for some $\psi \in A_n$ then $S(i) = \psi S(1)\psi^{-1}$, and the normal property of N gives $N \geqslant S(i)$. We observe that $S(2)$ is isomorphic to A_{n-1} in the natural way, and that $S(1) \cap S(2) \cong A_{n-2}$.

We now consider gp$\{S(1), S(2)\}$, which is contained in N. The order of this group is at least $|S(1)||S(2)||S(1) \cap S(2)|^{-1}$, as we see by counting the cosets of $S(1) \cap S(2)$. Thus

$$|N| \geqslant \{\tfrac{1}{2}(n-1)!\}\{\tfrac{1}{2}(n-1)!\}/\{\tfrac{1}{2}(n-2)!\} = \frac{n-1}{2}(n-1)!$$

But A_n has order $\frac{1}{2}n!$ Since $n-1$ does not divide n, we find that $|N| = |A_n|$. Thus $N = A_n$, and A_n is simple. □

COROLLARY 11.09. *A_X is simple if the set X is in one–one correspondence with the positive integers.*

Proof. We recall that A_X is a subgroup of R_X, each element of which moves only a *finite* number of symbols. Now A_X contains a subgroup isomorphic to A_n for each $n \geqslant 1$, namely the permutations affecting the subset $\{1,..., n\}$ of X only; we have $A_{n+1} \geqslant A_n$ for $n \geqslant 1$, and $A_X = \bigcup_{n \geqslant 1} A_n$.

Suppose that $1 < N \trianglelefteq A_X$. Then N contains an element $\rho \neq 1$ of some A_n. Further, $\rho \in A_{n+i}$ for each $i \geqslant 0$. But A_{n+i} is a simple group for $n+i \geqslant 5$, by the theorem, and $N \cap A_{n+i}$ is normal in A_{n+i}; so $A_{n+i} \leqslant N$. It follows that $A_X \leqslant N$. Therefore A_X is simple. □

We have made a simple choice of X in the above corollary, for the benefit of the reader unfamiliar with infinite set theory. However, this suffices to produce a group with a composition series and a subgroup with no composition series; namely A_X and its subgroup generated by

$$\{(123), (456), (789),...\}.$$

The details are left for the reader to complete.

Our final remarks on the examples of Chapter 1 concern Examples 1.28 and 1.29, multiplicative groups of non-singular matrices. One aspect of their importance is that a finite group of order n can always be exhibited as a suitable matrix group, according to Corollary 2.21. Our present remarks go no further than definitions.

A *field* K is a set in which two operations, called addition and multiplication, are defined; K is to be an abelian group with identity element 0 under addition, the non-zero elements of K are to form an abelian group under multiplication, and the familiar distributive law $a(b+c) = ab+ac$ is to be valid for all a, b, c in K. Familiar examples of fields are the real numbers, the rational numbers, and the residue classes modulo a prime (by Example 1.11).

We consider matrices with entries from K. A non-singular matrix has an inverse with entries from K, as may easily be shown, and it is natural to generalize Examples 1.28 and 1.29 by taking multiplicative groups of non-singular $n \times n$ matrices with entries from K.

Example 11.10. The set of all such matrices forms a group, the *general linear group* $GL(n, K)$. Detailed proof is easy and is omitted, in this and the following examples.

Example 11.11. The centre Z of $GL(n, K)$ consists of all matrices kI where k is a non-zero element of K. The proof of this statement is indicated in Example 4.26. The *projective general linear group* $PGL(n, K)$ is defined as $GL(n, K)/Z$.

Example 11.12. The *special linear group* $SL(n, K)$ is the subgroup of $GL(n, K)$ whose elements have determinant 1.

Example 11.13. The *projective special linear group* or *linear fractional group* $LF(n, K)$ is the central factor group of $SL(n, K)$. We state without proof that the groups $LF(n, K)$, with a few exceptions, are simple.

We now examine further examples, first fulfilling a promise made in Chapter 9 (just after Example 9.22).

Example 11.14. There is an abelian group A which is not the direct sum of its periodic subgroup and a torsionfree subgroup. Notice that A cannot be finitely generated, by Theorem 5.12.

Let p be any prime and let A_n be the cyclic group of order p^n, for each $n > 0$. The example is $A = \sum_{n>0}^{C} A_n$. Suppose that P is the periodic subgroup of A. We observe that P contains more than just $\sum_{n>0}^{D} A_n$; if $A_n = \text{gp}\{a_n\}$ then $f \in P$ where

$$nf = p^{n-1}a_n.$$

It will suffice for our purpose to show that A/P is not isomorphic to any subgroup of A.

Now A/P has the property that it contains certain non-trivial elements $a+P$ such that the equation

$$p^m(x+P) = a+P$$

has a solution $x+P$, for all $m > 0$. For instance, take $a \in A$ where

$$(2n)a = p^n a_{2n},$$
$$(2n-1)a = 1,$$

for $n = 1, 2, \ldots$, and fix m. If x in A is defined by

$$(2n)x = p^{n-m}a_{2n} \text{ for } n \geqslant m, \text{ and } 1 \text{ otherwise;}$$
$$(2n-1)x = 1,$$

for $n = 1, 2, \ldots$, then $p^m x - a \in P$, so that $p^m(x+P) = a+P$, as required.

On the other hand, if the element b of A is such that $p^m x = b$ has a solution for all $m > 0$, then $b = 1$. For if $b \neq 1$ then $nb \neq 1$ for some $n > 0$ while $ib = 1$ for $1 \leqslant i < n$. It follows that $p^{n+1}x = nb$ has a solution x in A_n, which contradicts one of two facts, namely $nb \neq 1$ and that $|A_n| = p^n$.

Therefore A/P is not isomorphic to any subgroup of A, and P cannot be a direct summand of A.

The remaining examples of this chapter require the concept of 'split extension'. Let G be a group with a normal subgroup N, and suppose that a transversal F of N in G can be chosen so that F is in fact a subgroup of G. This is a very special situation; one instance arises when $|G/N|$ and $|N|$ are finite and coprime, by Theorem 7.34. Each element of G can be expressed as $g_\alpha n$ where $g_\alpha \in F$, $n \in N$; and this expression is unique. Multiplication in G goes as follows:

$$(g_\alpha n_1)(g_\beta n_2) = (g_{\alpha\beta})(n_1^{g_\beta} n_2),$$

since $g_\alpha g_\beta = g_{\alpha\beta}$. There is a homomorphism from F into the automorphism group $A(N)$ of N, the images being defined by conjugation. Since $G/N \cong F$, we have a homomorphism from G/N into $A(N)$.

The object is to reverse the above procedure, to construct a group from given abstract groups F and N.

THEOREM 11.15. *Let groups F and N be given, together with a homomorphism ϕ from F into $A(N)$. Then there exists a group G containing a normal subgroup N' and a subgroup F' with the following properties:*

(i) $F' \cong F, \quad N' \cong N$;

(ii) $G/N' \cong F$;

(iii) *the automorphism of N' induced by conjugation with $x \in F'$ is the automorphism of N defined by $x\phi$, elements of F and F' and of N and N' being identified in the natural way.*

Proof. First we define a set G as follows:

$$G = \{(g_\alpha, n) : \alpha \in F,\ n \in N\}.$$

Here $\{g_\alpha\}$ is merely a set of symbols in one–one correspondence with $\{\alpha\}$, and G is a set of ordered pairs. Next we define multiplication in G by

$$(g_\alpha, n_1)(g_\beta, n_2) = (g_{\alpha\beta}, n_1^{g_\beta} n_2)$$

where $n_1^{g_\beta}$ denotes $n_1(\beta\phi)$; thus $\beta \in F$ and $\beta\phi \in A(N)$.

We now prove that G is a group. Closure is obvious. The associative law may be verified by a little calculation, left to the reader. The identity element is $(g_\epsilon, 1)$ where ϵ, 1 are the identity elements of F, N respectively. Finally, the inverse of (g_α, n) is $(g_\beta, (n^{-1})^{g_\beta})$ where $\beta = \alpha^{-1}$.

The following isomorphisms should be clear:

$$F \cong \{(g_\alpha, 1) : \alpha \in F\},$$
$$N \cong \{(g_\epsilon, n) : n \in N\}.$$

We therefore define F', N' in the obvious way, and (i) is complete. By now (ii) is obvious. As for (iii), we observe that

$$(g_\alpha, 1)^{-1}(g_\epsilon, n)(g_\alpha, 1) = (g_\epsilon, n^{g_\alpha})$$

follows from the definitions. But n^{g_α} was defined as $n(\alpha\phi)$. We therefore have (iii). □

The group G constructed in the theorem is called a *split extension of N by F*, or a *semidirect product of N by F*; it is perhaps better not to call it *the* split extension or semidirect product because it depends on ϕ as well as N and F. Its use, which we illustrate, is in the construction of particular groups.

Example 11.16. Consider again Example 1.27. This group G has a subgroup of index 2 isomorphic to R. It also contains an element ψ_0 which induces the inverting automorphism on R, and $\psi_0^2 = 1$; we discussed this earlier in Chapter 11. In fact it happens that G is a split extension of R by the group of order 2, and ϕ maps the generator of the latter onto the inverting automorphism.

The dihedral group of Example 1.25 may be similarly regarded as a split extension.

Example 11.17. We now use split extensions to construct a pair of important non-abelian groups of order p^3, where p is an arbitrary odd prime.

Let N_1 be the direct product of two groups of order p, with respective generators b, c. Let F_1 be of order p, and let ϕ_1 map a generator of F_1 onto α, the automorphism of N_1 specified by

$$b\alpha = bc, \quad c\alpha = c.$$

It is easy to see that this mapping does specify an automorphism α of order p. The resulting split extension G_1 may be presented as

$$\text{gp}\{a, b, c \colon a^p = 1,\ b^p = 1,\ c^p = 1,\ [b, c] = 1,\ b^a = bc,\ c^a = c\}$$

or, equivalently, as

$$\text{gp}\{a, b \colon a^p = 1,\ b^p = 1,\ [a, b]^p = 1,\ [a, b, a] = 1,\ [a, b, b] = 1\}.$$

The second construction starts by taking for N_2 the cyclic group of order p^2; let it be generated by b. Again F_2 is to have order p, and ϕ_2 is to map a generator onto the automorphism α of N_2 specified by

$$b\alpha = b^{1+p}.$$

After the details are tidied up, we obtain a group G_2 with presentation

$$\text{gp}\{a, b \colon a^p = 1,\ b^{p^2} = 1,\ b^a = b^{1+p}\}.$$

Example 11.18. Let N be cyclic of order 2^{n+1}, for any positive integer n, and let F be cyclic of order 4 with generator a. Further, ϕ is to be specified by $a\phi$ being the inverting automorphism of N. The split extension G which results has order 2^{n+3}.

Now the element $a^2b^{2^n}$ is central in G, if b generates N; put $Z = \text{gp}\{a^2b^{2^n}\}$. The factor group G/Z, of order 2^{n+2}, is of some interest. Since it coincides with the quaternion group of order 8 when $n = 1$, it is called the generalized quaternion group. It has the following presentation:

$$\text{gp}\{a, b \colon b^a = b^{-1},\ a^2 = b^{2^n}\}.$$

Next we discuss the holomorph of an arbitrary group G—this is a split extension of G by $A(G)$. Theorem 2.18 shows that the right regular representation of G is a subgroup of S_G isomorphic to G, where S_G is the symmetric group on the elements of G. Now $A(G)$ is also a subgroup of S_G, because any automorphism of G is a permutation of its elements. We have $\rho(G) \cap A(G) = 1$, where $\rho(G)$ denotes the copy of G in S_G, as may be seen from the actions of $\rho(G)$ and $A(G)$ on the element 1 of G. Before describing a split extension we need the following result.

THEOREM 11.19. *If $\phi \in A(G)$ then $\phi^{-1}\rho_g\phi = \rho_{g\phi}$, where ρ_g is the image of g in $\rho(G)$.*

Proof. Take any element x in G. Then

$$x(\phi^{-1}\rho_g\phi) = ((x\phi^{-1})\rho_g)\phi = ((x\phi^{-1})g)\phi = x(g\phi) = x\rho_{g\phi}.$$

The desired result follows. □

COROLLARY 11.20. *$A(G)$ normalizes $\rho(G)$ in S_G.* □

It follows that $\mathrm{gp}\{A(G), \rho(G)\} = A(G)\rho(G)$ and that $\rho(G)$ is normal in this. Indeed $A(G)\rho(G)$ is the split extension of $\rho(G)$ by $A(G)$, the automorphisms of $\rho(G)$ being specified as in Theorem 11.19. This group $A(G)\rho(G)$ is called the *holomorph* of G.

Example 11.21. Let G be cyclic of order p, an odd prime. Then $A(G)$ is cyclic of order $p-1$, and the holomorph of G has the following presentation:

$$\mathrm{gp}\{a, b\colon\ b^p = 1,\ a^{p-1} = 1,\ b^a = b^\alpha\},$$

where $[\alpha]$ has order $p-1$ modulo p (compare Theorem 11.05). Note that the holomorph contains the dihedral group of order $2p$ as a subgroup.

Example 11.22. Consider the automorphism group $A(Z)$ of the infinite cyclic group Z. If $Z = \mathrm{gp}\{z\}$ then there are only two elements which individually generate Z, namely z and z^{-1}. Since Z clearly has an automorphism interchanging these two, $A(Z)$ has order 2. It is now easy to give a presentation for the holomorph of Z:

$$\mathrm{gp}\{z, a\colon\ z^a = z^{-1},\ a^2 = 1\}.$$

The reader may be able to show that this group is isomorphic to that of Problem 5 of Chapter 3; it is the so-called 'infinite dihedral group'.

Problems

1. Let $\mathscr{M}$ be the set of all $m \times n$ matrices with entries from an abelian group A. Show that $\mathscr{M}$ forms a group under matrix addition (as in Example 1.32) and that $\mathscr{M}$ is isomorphic to the direct sum of mn copies of A.

2. Show that the two groups of order p^3 in Example 11.17 are non-isomorphic. Prove that there are no other non-abelian groups of that order, p being odd.

3. Prove that Q/Z is isomorphic to the direct sum of the groups $Z_{p_1}^{\infty}, \ldots, Z_{p_n}^{\infty}, \ldots$ where p_n is the nth prime.

*4. Find all groups of order 16.

5. (i) Prove that every finite group occurs as a subgroup of a suitable finite simple group.

(ii) Prove that every finite group can be represented as a (multiplicative) group of matrices of determinant 1.

6. Prove that the group of Example 1.23 is isomorphic to the group of 3×3 matrices with real entries and determinant 1.

7. An abelian group A is said to be *divisible* if the equation $nx = a$ has a solution x for each $n > 0$ and each $a \in A$. Which of the groups Z, Z_n, Z_p^∞, Q, R, Q/Z, R/Z are divisible?

8. Which of the following statements are true? Justify your choice.

(i) The symmetric group S_n has a unique subgroup of index 2.

(ii) Let p and q be primes with $p > q$. There is precisely one non-abelian group of order pq if and only if $p \equiv 1$ modulo q.

(iii) Let p_n denote the nth prime and Z_n the group of that order. Then

$$A\Big(\prod_{n>0}^C Z_n\Big) \cong \prod_{n>0}^C A(Z_n).$$

(iv) If $G = \prod_{n>0}^D G_n$ where each G_n is a characteristic subgroup of G, then $A(G) \cong \prod_{n>0}^D A(G_n)$.

(v) All groups of (finite) order m are abelian if and only if m is divisible by the square of no prime.

***9. Let K denote the field of residue classes with 5 elements.

(i) Prove that the centre Z of $SL(2, K)$ has order 2.

(ii) Prove that Z is the only proper normal subgroup of $SL(2, K)$.

(iii) Deduce that $G = \delta(G)$ when $G = SL(2, K)$.

(iv) What is the Sylow 2-subgroup of $SL(2, K)$, as abstract group?

(v) Give an alternative description of $LF(2, K)$.

10. Let G be a finite group. Show that, in the regular representation of G, any two isomorphic subgroups of G map onto conjugate subgroups of S_G.

11. Show that $\quad \text{gp}\{a, b : a^2 = b^2 = (ab)^4 = 1\}$

is a presentation of the dihedral group of order 8, and that this group G has an automorphism α of order 2 which interchanges a and b.

Let H be the split extension of G by a group of order 2, with α playing the obvious role; thus H has order 16. Is H a dihedral group, or a generalized quaternion group, or neither?

*12. Prove that

$$\text{gp}\{a, b : a^b = a^{31},\ a^{10} = b^{10},\ a^{100} = 1\}$$

has order 10^3, and identify its Sylow 2-subgroup.

Prove further that none of the three given relations is redundant, but that the same group may be presented as follows:

$$\mathrm{gp}\{c, d : c^{10} = d^{10} = [c, d]\}.$$

*13. A hamiltonian group is a non-abelian group in which every subgroup is normal. Let G be a hamiltonian group.

(i) Prove that $[x, y, y] = 1$ for all x, y in G.
(ii) Prove that G is periodic.
(iii) Deduce that G is the direct product of nilpotent p-groups for various primes p.
(iv) Prove that the Sylow p-subgroups, for odd primes p, are abelian.
(v) Prove that the Sylow 2-subgroup G_2 contains a quaternion group Q_8; that the centralizer C of Q_8 in G_2 has exponent 2; and that $G_2 = CQ_8$.

Deduce that a finite group is hamiltonian if and only if it is the direct product of a quaternion group, an abelian group of exponent 2, and an abelian group of odd order.

***14. Let G be a finite group of exponent 3, and generated by $\{a, b, c\}$. Assuming for the moment that G is nilpotent of class 3 but not of class 2, prove the following facts about G:

(i) $\delta(G)$ has order 3^4 and is elementary abelian;
(ii) $G/\delta(G)$ has order 3^3 and is elementary abelian;
(iii) $\gamma_3(G)$ has order 3.

Now try to construct a group H of order 3^7, exponent 3, and class 3 but not 2, as follows. Take an elementary abelian group of order 3^4 and form a split extension H_1 of this by a group of order 3; form a similar split extension H_2 of H_1, and a third split extension H of H_2. At each stage use information gained earlier about the structure of G.

Revision problems

1. Distinguish the true statements among the following from the false, giving appropriate proofs.

(i) a and b are conjugate in G if and only if $a^{-1}b \in \delta(G)$.

(ii) The non-empty complex C in the group G is a subgroup if and only if $Cx = Cy$ implies $x = y$.

(iii) There is a group G such that $G/\zeta(G)$ is isomorphic to the additive group Q of rationals.

(iv) If $X = \begin{pmatrix} 1 & 2 \\ -1 & -2 \end{pmatrix}$ then the distinct positive powers of X form a group.

(v) A group generated by an arbitrary distinct pair of its non-trivial elements is cyclic.

(vi) Every group of order n^2 is abelian (for each positive integer n).

(vii) Any subgroup of order 2 in a group is central.

(viii) If a, b are elements of the group G then the orders of the elements $a^{-1}b$, ab^{-1}, ba^{-1}, $b^{-1}a$ are all equal.

(ix) All cyclic groups of infinite order are isomorphic, and therefore each is generated by any one of its elements.

(x) If a, b are any elements of any group then $a^b \in \text{gp}\{a\}$.

2. A central product of groups G, H with isomorphic centres $\zeta(G)$, $\zeta(H)$ is defined as follows: it is $(G \times H)/N$ where N consists of all pairs (x, y) with $x \in \zeta(G)$, $y \in \zeta(H)$, and x, y corresponding under the given isomorphism.

(i) Prove that the central product is well defined if $\zeta(G)$, $\zeta(H)$ have order 2.

(ii) Prove that the central product of two quaternion groups is isomorphic to that of two dihedral groups of order 8.

(iii) Is this group isomorphic to the central product of one quaternion group and one dihedral group?

3. Are the following groups isomorphic?

$$\text{gp}\{a, b : a^8 = 1, \quad a^4 = b^4, \quad a^b = a^{-1}\},$$

$$\text{gp}\{a, b : a^8 = 1, \quad a^4 = b^4, \quad a^b = a^3\}.$$

4. The subgroup B of the abelian group A is said to be *pure* if the equation $nx = b$ (n positive integer, $b \in B$) has a solution x in B whenever it has a solution in A.

Prove that a subgroup of a finitely generated torsionfree abelian group is a direct summand if and only if it is pure.

Is the corresponding result true for subgroups of finitely generated abelian groups?

5. Prove that it is impossible for $G/\zeta(G)$ to have a normal subgroup isomorphic to the quaternion group, whatever group G is.

Deduce that $G/\zeta(G)$ has no hamiltonian normal subgroup.

6. Prove that if G is any group and $N \trianglelefteq G$ then $\gamma_n(G/N) = \gamma_n(G)N/N$ for $n \geqslant 1$.

Deduce that if F is a free group then $F/\gamma_{n+1}(F)$ is nilpotent of class n.

Hence (or otherwise) show that a free basis for F cannot contain two distinct conjugate elements.

7. Prove that the automorphism group of the dihedral group of order $2n$ (see Example 1.25 on page 11) is isomorphic to the holomorph of the cyclic group of order n, provided $n \geqslant 3$.

8. Let a, b be fixed elements of the group G, and n be a positive integer.

Prove that $$[a, b]^n = (a^{-1}b^{-1})^n(ab)^n c_1 \dots c_{n-1}$$

for suitable commutators $c_1, \dots, c_{n-1}$. Prove further that if $|G/\zeta(G)| = n$ then $(ab)^n = (ba)^n$, and that $[a, b]^n$ may be expressed as the product of some $n-1$ commutators. Deduce that if $G/\zeta(G)$ is finite then so is $\delta(G)$.

9. The group G contains a finite normal subgroup H in which lies a Sylow p-subgroup G_p of G. Prove that $G = HN_G\{N_H(G_p)\}$ where N_G, N_H denote normalizers in G, H respectively.

10. The normal subgroups A, B of the group G are nilpotent of classes c_1, c_2 respectively. Prove that AB is nilpotent of class $c_1 + c_2$.

11. Let G be a group and γ a mapping of G into G. For each x in G define $[x, \gamma]$ to be $x^{-1}(x\gamma)$ and $[\gamma, x]$ to be $[x, \gamma]^{-1}$. Let a, b be two fixed elements of G.

Prove that $$[a, b]\gamma = [a\gamma, b\gamma]$$

if and only if

$$[[b, a], [\gamma, b]][[b, a], \gamma]^{[\gamma, b]}[\gamma, b, a\gamma][a, \gamma, b^a] = 1.$$

12. The group G is a finite 2-group in which $G/\delta(G)$ has order 4.

(i) Prove (by induction, for instance) that G contains a cyclic subgroup of index 2.

(ii) Show that (i) is false if $\delta(G)$ is replaced by $\Phi(G)$.

(iii) Show that (i) is false if the prime 2 is replaced by any other prime.

13. Let G be a finite p-group. Prove that if $\Phi(G)$ has centre of order p and is non-abelian then $\zeta(G) \cap \Phi(G) = \zeta\{\Phi(G)\}$; and deduce that such a group G does not exist. Hence prove:

(i) All groups of order p^5 are metabelian.

(ii) All groups of order 2^6 are metabelian (use the previous problem).

14. The element g of the group G has finite order and lies in a finite conjugacy class. Prove that g lies in a finite normal subgroup of G.

Deduce that no group contains an even number of elements of order 2.

15. A subgroup H of the group G is said to be *quasinormal* if H commutes with every subgroup of G.

Let $G = A \times B$, where

$$A = \text{gp}\{a, b : a^b = a^{1+p},\ b^p = 1\},$$
$$B = \text{gp}\{c : c^{p^2} = 1\},$$

and p is an odd prime. Find two quasinormal subgroups of G whose intersection is not quasinormal.

16. The subgroups A, B of the group G are such that $AB = BA$. Prove that A and B are not conjugate in AB unless $A = B$.

Deduce that if a group G has a maximal subgroup M which is quasinormal in G (see the previous problem) then $M \trianglelefteq G$.

Show that any subgroup of index p in the finite group G is normal if p is the smallest prime dividing $|G|$.

17. Prove that a p-group of finite exponent is nilpotent if it contains a nilpotent subgroup of finite index.

18. The group G has $G/\delta_1(G)$ a cyclic p-group (p prime). Prove that no normal subgroup of G has index a positive power of p. Deduce from this that if $\delta_2(G)/\delta_3(G)$ is a finite cyclic p-group then $\delta_2(G) = \delta_3(G)$.

19. Let G be a group in which, for each n, $\delta(G)$ contains an element not in any way expressible as the product of n commutators. Let $G_n \cong G$ for $n = 1, 2, \ldots$ and put $H = \prod_{n>0}^{C} G_n$. Show that there exists an element of $\prod_{n>0}^{C} \delta(G_n)$ which does not lie in $\delta(H)$.

Is it true that $\zeta(H) = \prod_{n>0}^{C} \zeta(G_n)$?

20. In the arbitrary group G, the subgroup H_3 is generated by all elements in G which do *not* have order 3. Let $h \in H_3$, $x \notin H_3$, and prove:

(i) $hh^x h^{x^2} = 1$;

(ii) $[h^x, h^y] = 1$ if $xH_3 \neq yH_3$;

(iii) $H_3 = 1$, unless H_3 has index 3 or 1 in G.

Deduce that a finite (non-trivial) group G has exponent 3 if and only if every element outside $\Phi(G)$ has order 3.

21. Let the group G have a fixed finite set $\{g_1,\ldots,g_n\}$ of generators, and consider subgroups of fixed finite index k in G. Such a subgroup K determines an equivalence relation in G in the usual way: $x \sim y$ means $xy^{-1} \in K$. Show that if $\{h_1,\ldots,h_k\} \subseteq \{g_1^{\pm 1},\ldots,g_n^{\pm 1}\}$ then two of the following elements of G are equivalent:

$$1, \quad h_1, \quad h_1h_2, \quad \ldots, \quad h_1h_2\ldots h_k.$$

Deduce that G has only a finite number of subgroups of index k.

22. Use the general method of the previous problem to show that a subgroup of finite index in a finitely generated group is itself finitely generated.

23. Prove that a finitely generated periodic soluble group is finite. (Hint: use induction on the solubility length and apply the result of the previous problem.)

24. Prove that if the group G is nilpotent of class 2^n-1, where n is a positive integer, then G is soluble of length n.

25. Let H_1, H_2 be given groups and let K be a subgroup of $G = H_1 \times H_2$. Define subsets F_1, F_2 of H_1, H_2 as follows:

$$F_1 = \{h_1 : (h_1, h_2) \in K \text{ for some } h_2 \in H_2\},$$

$$F_2 = \{h_2 : (h_1, h_2) \in K \text{ for some } h_1 \in H_1\}.$$

(i) Prove that $F_i \leqslant H_i$ for $i = 1, 2$.
(ii) Prove that if $K_i = K \cap H_i$ then $K_i \trianglelefteq F_i$ for $i = 1, 2$.
(iii) Prove that $K/K_1 \cong F_2$, $K/K_2 \cong F_1$.
(iv) Prove that $F_1/K_1 \cong F_2/K_2$; and that $(h_1, h_2) \in K$ if and only if the images of h_i in F_i/K_i $(i = 1, 2)$ correspond in this isomorphism.

26. Let G be a finite p-group on two generators, having exponent p^2 and containing an element of order p^2. Show that the following statements are equivalent:
(i) Every element of order p^2 lies in $\Phi(G)$.
(ii) $|G/H(G)| > p$, where $H(G)$ is the subgroup generated by the elements of order p^2.

27. The abelian group A has subgroups $A_1,\ldots, A_n$ and homomorphisms ϕ_i from A to A_i, and ψ_i from A_i to A, satisfying the following conditions for $1 \leqslant i \leqslant n$:
(i) $\psi_i\phi_i$ is the identity mapping on A_i;
(ii) $\psi_i\phi_j$ maps A_i onto 0 if $j \neq i$.

Prove that $A \cong A_1 \oplus \ldots \oplus A_n$ if and only if $\sum_{i=}^{n} a\phi_i\psi_i = a$ for each element a of A.

28. The finite group G has a subgroup H for which $|G:H|$ is a power of the prime p. Show that if G_p is a Sylow p-subgroup of G then $H \cap G_p$ is a Sylow p-subgroup of H, and $G = HG_p$.

Prove that the following conditions on the element x of p-power order of the finite group G are equivalent:

(i) The least normal subgroup of G containing x is a p-group.

(ii) There is a normal subgroup N of G containing x, and x has n conjugates as an element of N where n is some power of p.

29. The finite group G has the property that no maximal subgroup is normal. Prove that either G is abelian or two distinct maximal subgroups have non-trivial intersection, by assuming that the latter alternative is false and counting the elements of G in each conjugacy class of maximal subgroups.

30. Suppose that G is a finite group in which every proper subgroup is nilpotent. Prove that if G is simple then any two distinct maximal subgroups have trivial intersection. Deduce from this and the result of the last problem that G must in fact be soluble.

31. Prove the theorem that a finite non-abelian group G contains a pair of distinct conjugate elements which commute, by means of the following steps.

(i) Reduce the problem to the case in which every maximal subgroup of G is abelian but not normal.

(ii) Prove that $M_1 \cap M_2 = \zeta(G)$, where M_1, M_2 are distinct maximal subgroups.

(iii) Apply Problem 29 to investigate $G/\zeta(G)$.

*32. Let H be a finitely generated subgroup of G, which is defined to be $\prod^C_{\lambda\in\Lambda} F_\lambda$, each F_λ being isomorphic to the finite group F. Prove that, for fixed λ, the mapping from H to F_λ which sends h to λh is a homomorphism and that its kernel K_λ has finite index in H. Deduce that H is finite.

Prove further that if each F_λ is finite and $|F_\lambda|$ is bounded independently of λ then every finitely generated subgroup of $\prod^C_{\lambda\in\Lambda} F_\lambda$ is finite.

*33. Show that the following statements are equivalent:

(i) Every residually finite group having d generators and (finite) exponent e is finite.

(ii) The finite groups with d generators and exponent e have their orders bounded by a function of d and e.

34. Let n be the number of distinct primes dividing the positive integer m. Find a set of n elements which generates the cyclic group Z_m of order m, such that no proper subset generates Z_m. Find a set of n defining relations in such a set of generators, with the property that no relation is redundant.

35. Prove that $\mathrm{gp}\{a,b,c:b^c = b^3,\ c^a = c^3,\ a^b = a^3\}$ is a finite 2-group which cannot be generated by any two of its elements. What is the order of the group?

Prove that $\mathrm{gp}\{a,b,c : b^c = b^{-2}, c^a = c^{-2}, a^b = a^{-2}\}$ is a finite 3-group which cannot be generated by any two of its elements.

(Hint: Lemma 9.29 on page 189 may be of use.)

36. The group G is defined by generators a, b, c and relations

$$b^c = b^3, \quad c^a = c^{-1}, \quad a^b = a^{-1},$$

$$a^4 = b^4 = c^4, \quad a^8 = 1.$$

(i) Prove that G has order 2^7.

(ii) Prove that if $x, y \notin \Phi(G)$ then $\mathrm{gp}\{x\} \cap \mathrm{gp}\{y\} \neq 1$.

37. (i) Prove that a finite group G is nilpotent if and only if $G/\Phi(G)$ is nilpotent.

(ii) Prove that, if G is a finite group with $H \leqslant G$, $N \trianglelefteq G$, and $N \leqslant \Phi(H)$, then $N \leqslant \Phi(G)$.

(iii) Prove that, if G is a finite group with a normal subgroup H such that both H and $G/\delta(H)$ are nilpotent, then G is nilpotent.

*38. The *upper p-series*

$$1 = P_0 \leqslant N_0 < P_1 < \ldots < P_r \leqslant N_r = G$$

of the finite group G is defined as follows: p is any prime, N_i/P_i is the largest normal subgroup of G/P_i with order prime to p for $0 \leqslant i \leqslant r$, and P_{i+1}/N_i is the largest normal p-subgroup of G/N_i for $0 \leqslant i < r$. The group G is said to have *p-length r*.

(i) Prove that the upper p-series is a well-defined invariant series for G.

(ii) Prove that, if G is soluble, then P_1 contains the centralizer C of P_1/N_0. (C is $\{x : x \in G, [g,x] \in N_0$ for all $g \in P_1\}$.) (Hint: the Schur–Zassenhaus theorem may be useful.)

(iii) Prove that a finite soluble group with abelian Sylow p-subgroups has p-length 1.

*39. The *socle* $\sigma(G)$ of a finite group G is defined to be the subgroup generated by all the minimal normal subgroups of G. Prove that $\sigma(G)$ is in fact the direct product of a selection of these normal subgroups.

Let $N \leqslant \sigma(G)$, and $N \trianglelefteq G$. Prove that N is also the direct product of a suitable set of minimal normal subgroups of G. (Hint: Problem 25 may be helpful.)

Deduce that a finite group with no proper characteristic subgroup is the direct product of isomorphic simple groups.

**40. An important property of a finite soluble group G whose Sylow subgroups are all abelian is that $\delta(G) \cap \zeta(G) = 1$. Prove this in the steps indicated below, or otherwise.

(i) Reduce the problem to the case in which $\delta(G) \cap \zeta(G)$ is non-trivial and is the unique minimal normal subgroup of G. Consider this case only in what follows.

(ii) Show that G is metabelian, and that $\delta(G)$ is a p-group. Deduce that $G = G_p G_\pi$ where G_p is a Sylow p-subgroup and G_π is a suitable abelian Hall π-subgroup.

(iii) The elements of order p in $\delta(G)$ form a subgroup P of the form $P_1 \times P_2$, where $P_1 = \delta(G) \cap \zeta(G)$ and P_2 is a suitable subgroup. Each element of G_π induces an automorphism θ of P which fixes P_1. If $x \in P$ then $x = x_1 x_2$ ($x_i \in P_i$ for $i = 1, 2$) and we may form $x_3 = \prod_\theta \{(x\theta)_2 \theta^{-1}\}$.

Prove that the set of all such x_3 is a normal subgroup of G, and deduce that $P = \delta(G) \cap \zeta(G)$.

(iv) Show that G cannot exist.

*41. We define a *subcartesian product* H of the groups $A_1,\ldots, A_n,\ldots$ as follows. Let $G = \prod^C_{n>0} A_n$ and let $H \leqslant G$; then we require that

$$\{n\phi : \phi \in H\} = A_n$$

for each $n > 0$.

Now let A be a fixed group and B be a subgroup of A. Take $A_n \cong A$ for each n. Show that the constant functions on $\{1,\ldots, n,\ldots\}$ generate a subgroup A' of G isomorphic to A, and that A' contains a subgroup B' isomorphic to B.

Prove that B' and $\prod^D_{n>0} A_n$ generate in G a subcartesian product of $A_1,\ldots, A_n,\ldots$.

Deduce that B may be obtained from A by means of these operations:

(i) taking a subcartesian product S of isomorphic copies of A;

(ii) taking a factor group of S.

*42. Let F be a free group with free basis $\{a, b\}$, and suppose that F contains elements u, v for which

$$[a, b]^2 = [u, v].$$

Show that even if u, v are reduced words then some cancellation occurs when $u^{-1}v^{-1}uv$ is reduced.

Let u, v be chosen so that the sum of their lengths is minimal. Prove that

(i) none of u^{-1}, v^{-1}, u, v cancels completely in an adjoining word; and

(ii) u and v are not of the form $x^{-1}wx$ unless $x = 1$.

Deduce that the cancellation in $[u, v]$ occurs either between v^{-1} and u only; or between u^{-1} and v^{-1}, or u and v, or both.

Hence show that $[a, b]^2$ is not a commutator in F.

*43. Prove that $\text{gp}\{a, b : a^2 = b^3 = (ab)^3 = 1\}$ is the alternating group on 4 symbols.

The finitely generated group G contains an element of each order 1, 2, 3 but no elements of other orders. Prove the following statements about G:

(i) Two elements of order 2 generate S_3 or Z_2 (of order 2) or $Z_2 \times Z_2$.

(ii) No element of order 2 lies in both S_3 and $Z_2 \times Z_2$.

(iii) G does not contain both S_3 and $Z_2 \times Z_2$.

(iv) G does not contain both $Z_2 \times Z_2$ and $Z_3 \times Z_3$ (where Z_3 is the group of order 3).

(v) G is finite, and contains either a normal subgroup of exponent 2 with index 3 or a normal abelian subgroup of exponent 3 with index 2.

44. Let G be a nilpotent group with a normal non-abelian subgroup H such that $\zeta(H)$ is cyclic of prime or infinite order. Prove that $\zeta(H) \leqslant \zeta(G)$.

Let g be such that $\zeta(H)g$ is a non-trivial central element of $G/\zeta(H)$ and consider the mapping ϕ defined by $x\phi = [x, g]$ for all $x \in G$. Show that ϕ is a homomorphism and deduce (from examination of the kernel of ϕ) that H cannot lie in $\Phi(G)$.

**45. Prove that if G is the set-theoretic union of its subgroups $G_1, \ldots, G_n$, but not of any proper subset of them, then $G_1 \cap \ldots \cap G_n$ has finite index in G.

46. Let G denote the group defined by generators a, b, c and relations

$$a^{2^n} = b^4 = c^2 = 1,$$

$$a^b = a^{1+2^{n-1}}, \quad b^c = b^{-1}, \quad [c, a] = 1,$$

where $n \geqslant 3$. Prove the following facts about the full group of automorphisms of G:

(i) it is a 2-group;

(ii) it is abelian.

***47. In any group G the set of elements $\{a : [a, x, x] = 1 \text{ for all } x \text{ in } G\}$ is denoted by $E_2(G)$. Prove the following facts about $E_2(G)$:

(i) If $a \in E_2(G)$ then a lies in an abelian normal subgroup of G, and $a^{-1} \in E_2(G)$.

(ii) Expand the relation $[a, xy, xy] = 1$ to prove

$$[a, x, y][a, y, x][a, x, y, x] = 1$$

for all x, y in G.

(iii) By replacing x by x^{-1} in (ii), show that

$$[a, x, y][a, y, x] = 1,$$

$$[a, x, y, x] = 1.$$

(iv) Deduce that $E_2(G) \leqslant G$.

(v) By making use of Lemma 9.29, prove that

$$[a, x, y]^{-2} = [x, y, a].$$

(vi) Prove that $[a, x, y, z]^4 = 1$ for all x, y, z in G; and deduce that $E_2(G) \leqslant \zeta_3(G)$ provided no element of G has order 2.

**48. Let G be the (multiplicative) group generated by a, b where

$$a = \begin{pmatrix} 12 & 3 \\ 14 & 5 \end{pmatrix}, \qquad b = \begin{pmatrix} 7 & 6 \\ 7 & 11 \end{pmatrix},$$

and the entries are residue classes modulo 17. Prove that

$$a^2 = b^3 = (ab)^4.$$

Hence (or otherwise) show that G is a soluble group of order 48, with $\zeta(G)$ of order 2 and $G/\zeta(G) \cong S_4$. Is G isomorphic to $GL(2, K)$ where K is a field with 3 elements?

**49. In the group G the derived group $\delta(G)$ is infinite cyclic and $\delta(G) \leqslant \zeta(G)$. Prove that $\delta(G)$ is generated by one commutator.

50. Give a critical discussion of the following statements.

(i) The definition of a group is when you have a set and it also has a multiplication table.

(ii) The group of all non-singular 2×2 matrices is an isomorphic group.

(iii) All groups of prime order are cyclic, and since 13 and 17 are prime all groups of those orders are cyclic and so abelian. But clearly 15 is the mean of 13 and 17, and it follows that any group of order 15 is abelian.

(iv) Let the group G have a subgroup H of index 2. The set of elements not in H is a coset but not a subgroup, so H is the only subgroup of G having index 2. Therefore H is a characteristic subgroup of G.

(v) Let S be a subgroup of the free group F. If $w = 1$ where w is a reduced word in S, then the freeness of F shows that w is the empty word. Therefore S is free.

Appendix

Elementary set and number theory

THE purpose of this survey of basic concepts does not comprise formal definition of sets, mappings, numbers, and the like. We are more concerned with settling genuine queries and doubts that might appear in the mind of the absolute beginner, as well as with describing notation and conventions.

Sets of numbers are the most familiar and elementary examples of sets. The elements of the set Z of integers are $0, \pm 1, \pm 2, \ldots$, and we write

$$Z = \{0, \pm 1, \pm 2, \ldots\}.$$

The set of positive integers is $\{1, 2, \ldots\}$. The set Q of rational numbers is described in our set notation as

$$Q = \left\{x \colon x = \frac{p}{q}, \quad p \in Z, \quad q \in Z, \quad q \neq 0\right\}.$$

The symbols following the colon give the defining property of Q, and '$p \in Z$' means that p is an element of the set Z, that is p is an integer. Other fundamental number systems are the set of real numbers and the set of complex numbers.

If A is a set and if each element of the set B belongs to A, then we say that B is a subset of A and write $B \subset A$. If there is an element x such that $x \in A$ and $x \notin B$, then we say that B is a proper subset of A and write $B \subset A$. Thus we have $Z \subset Q$, for example. Note that $A = B$ if and only if $A \subseteq B$ and $B \subseteq A$ are both valid statements.

The best attitude towards the empty set $\emptyset$ is, perhaps, to regard it as an interesting curiosity, a convenient fiction. To say that $x \in \emptyset$ simply means that x does not exist. Note that it is conventionally agreed that $\emptyset$ is a subset of every set, for elements of $\emptyset$ are supposed to possess every property.

Union and intersection of sets A, B are defined in the standard way

$$A \cup B = \{x : x \in A \text{ or } x \in B\},$$

$$A \cap B = \{x : x \in A \text{ and } x \in B\},$$

respectively. Since $\emptyset$ is a set, $A \cap B$ is always a set. If A is the set of three-sided polygons and B is the set of four-sided polygons then it is clear that $A \cap B = \emptyset$.

The reader should be at ease with the idea of a set of sets, that is a set whose elements are themselves sets. Suppose A is some fixed set. Then

$$\{x : x \text{ is a subset of } A\}$$

is a perfectly respectable set of sets. (Factor groups are sets of sets.)

A partition of a set A is another important example of a set of sets. It is a set P whose elements are non-empty subsets of A such that each element of A lies in one and only one element of P. A possible partition of Z is $\{Z_1, Z_2\}$ where Z_1, Z_2 are the subsets of even, odd integers respectively.

We must also mention sets of ordered pairs. If A, B are sets and if $a \in A$, $b \in B$, we can form the pair $\{a, b\}$; it is simply a set with two elements. We may think of a, b as occurring in the order 'first a, then b', and in so doing form the concept of an ordered pair, conventionally written as (a, b) in this case. We can clearly form sets whose elements are ordered pairs; for instance

$$A \times B = \{(a, b) : a \in A,\ b \in B\}.$$

This can readily be generalized. Given n sets $A_1, \ldots, A_n$ we can consider n-tuples, or ordered sets of n elements; an n-tuple is written as $(a_1, \ldots, a_n)$ where $a_i \in A_i$ for $1 \leqslant i \leqslant n$, and the set of all n-tuples is denoted by $A_1 \times \ldots \times A_n$ in these circumstances.

Next we consider mappings from a set A into a set B. A mapping ϕ associates with each element x of A precisely one element y of B, a fact described in the equation $y = x\phi$. Note that we write the mapping symbol ϕ 'on the right', which is the usual convention in algebra. Two mappings ϕ and ψ from A into B are equal if and only if $x\phi = x\psi$ for all $x \in A$.

We say that ϕ is *onto* B, or simply *onto*, if $b = x\phi$ has at least one solution x in A for each element b in B; thus each element b in B appears as the image of a suitable element of A. We say that ϕ is *one–one* if $x\phi = y\phi$ implies $x = y$; thus distinct elements of A have distinct images in B. The case in which ϕ is a one–one mapping from A onto B is of special interest; ϕ is called a one–one correspondence. A set is called *countable* if it is finite or is in one–one correspondence with the set Z of integers.

We turn to mappings of rather special sets. Suppose that A is some set and ϕ is a mapping of $A \times A$ into another set B; then ϕ is called a binary operation on A. All that ϕ does is to map each ordered pair (a_1, a_2), where $a_1 \in A$ and $a_2 \in A$, onto an element of B. If ϕ maps $A \times A$ into A then we say that 'A is closed under the binary operation ϕ'. (Group 'multiplication' is of course a binary operation on the group.)

As far as this book is concerned there is no essential difference in the meanings of the terms mapping, function, and operation.

Suppose next that ϕ is a mapping from A into B, and ψ is a mapping from B into C, where A, B, C are certain sets. Then we can define a mapping from A into C by applying first ϕ and then ψ, obtaining $(x\phi)\psi$ from the element x of A. This mapping is called the product of ϕ and ψ, and is written $\phi\psi$. The definition and notation yield the equation

$$x(\phi\psi) = (x\phi)\psi.$$

The reader will probably find it easy to persuade himself that $\phi\psi$ is onto only if ψ is onto, and that $\phi\psi$ is one–one only if ϕ is one–one; the product of two one–one correspondences is also a one–one correspondence.

Another interesting exercise arises when we have mappings ϕ_1, ϕ_2 from A into B, and mappings ψ_1, ψ_2 from B into C. The reader is invited to prove that ϕ is onto if and only if $\phi\psi_1 = \phi\psi_2$ implies $\psi_1 = \psi_2$, and that ψ is one–one if and only if $\phi_1\psi = \phi_2\psi$ implies $\phi_1 = \phi_2$; the mappings ϕ_1, ϕ_2, ψ_1, ψ_2 are to be regarded as arbitrary.

A most important property of the multiplication of mappings that has just been described is its so-called associative property. Let A, B, C, D be sets and let ϕ, ψ, ω be mappings from A to B, from B to C, from C to D respectively. The property in question is expressed by the equation

$$\phi(\psi\omega) = (\phi\psi)\omega.$$

In terms of an arbitrary element x of A, this is just

$$(x\phi)(\psi\omega) = \{x(\phi\psi)\}\omega;$$

the result of acting on x first with ϕ and then with $\psi\omega$ coincides with the result of acting first with $\phi\psi$ and then with ω; the products $\psi\omega$, $\phi\psi$ are defined as indicated above.

The associative law is intuitively very obvious. Its truth should be manifest once its meaning is grasped. We give a formal account of its proof for the sake of completeness, rather than with an intention of thereby increasing understanding.

The proof is just a calculation based on the product definition. The meaning of $\psi\omega$ shows that

$$(x\phi)(\psi\omega) = \{(x\phi)\psi\}\omega,$$

while the meaning of $\phi\psi$ shows that

$$x(\phi\psi) = (x\phi)\psi$$

and hence $$\{x(\phi\psi)\}\omega = \{(x\phi)\psi\}\omega.$$

We therefore have $$(x\phi)(\psi\omega) = \{x(\phi\psi)\}\omega,$$

both sides being equal to $\{(x\phi)\psi\}\omega$. Since x was an arbitrary element of A we have the desired equation $\phi(\psi\omega) = (\phi\psi)\omega$.

Note the special case of this result in which $A = B = C = D$, so that ϕ, ψ, ω are mappings from A into A. In that case products like $\psi\omega$, $\phi(\psi\omega)$, etc., are of course also mappings from A into A.

The identity mapping ι of a set A onto itself is defined by $x\iota = x$ for all x in A. It may perhaps not seem natural to regard this as a mapping until one is convinced of its utility (this sort of feeling may arise from the set $\emptyset$, in another context).

Next, let ϕ be a mapping from A to B and ψ be a mapping from B to A, and suppose that $\phi\psi$ is the identity mapping (on A); then we say that ψ is a right inverse of ϕ, and ϕ is a left inverse of ψ. If, on considering $\phi\psi$ and $\psi\phi$, we find that ψ is both a right inverse and a left inverse of ϕ then we call ψ an inverse of ϕ and write $\psi = \phi^{-1}$. The reader may like to find examples of right or left inverses that are not inverses.

In fact ϕ has a right inverse if and only if ϕ is one–one, and ϕ has a left inverse if and only if ϕ is onto; no proof of these statements will be attempted. They imply that ϕ has an inverse if and only if ϕ is a one–one correspondence; this is almost trivial to verify independently.

The next concept we consider is that of a one–one mapping of a set A onto itself; this sort of mapping is called a permutation of A. It is nothing but a one–one correspondence of A with itself. Clearly a permutation has an inverse, and clearly this inverse is also a permutation. We note that, in the special case when A is a finite set, a mapping ϕ from A into A is onto if and only if it is one–one; so either of these conditions is equivalent to ϕ being a permutation of A.

Suppose A is a finite set. If A has only one element then the only permutation of A is the identity mapping. How many permutations does A have when it contains N elements? Suppose that there are precisely $p(n)$ one–one mappings from one set A_n with n elements onto another set B_n also containing n elements, for each $n \geqslant 1$; clearly $p(1) = 1$. Let $n > 1$. A definite element x of A_n may be mapped onto any one of the elements of B_n; therefore the image of x may be chosen in n ways. A one–one mapping of A_n onto B_n may then be completed by mapping the other $n-1$ elements of A onto the remaining $n-1$ elements of B_n, in a one–one manner, and this may be effected in precisely $p(n-1)$ ways. Therefore $p(n) = np(n-1)$ for $n > 1$. An easy induction shows that $p(n) = n!$, the well-known factorial function. Now take $B_n = A_n$. The conclusion is that there are $N!$ permutations on a set of N elements.

Special notation is needed to obtain such properties of permutations as we require. It is traditional, but by no means essential, to regard permutations as acting on a set whose elements are labelled with distinct positive integers, if the set is finite; the positive integers or all the integers may be used for countable sets. We note that these numbers appear only as markers or labels, their additive and multiplicative properties are not required.

Suppose then that ρ is a permutation of the set $\{1, 2, \ldots, n\}$. The standard notation for ρ is

$$\begin{pmatrix} 1 & 2 & \ldots & n \\ 1\rho & 2\rho & \ldots & n\rho \end{pmatrix}. \tag{1}$$

An abbreviated notation is common and useful. Consider the integers $1,\ 1\rho,\ 1\rho^2,\ldots$. We must come to an integer m, where $1 \leqslant m \leqslant n$, such that $1\rho^m = 1$; for we must have a repetition of integers some time (they number only n) and if $1\rho^{m_1} = 1\rho^{m_2}$ with $0 \leqslant m_1 < m_2$ then $1\rho^{m_2-m_1} = 1$. The action of ρ on $\{1, 1\rho,\ldots, 1\rho^{m-1}\}$ is officially given as

$$\begin{pmatrix} 1 & 1\rho & \ldots & 1\rho^{m-1} \\ 1\rho & 1\rho^2 & \ldots & 1 \end{pmatrix}, \tag{2}$$

but we often abbreviate this to $(1, 1\rho,\ldots, 1\rho^{m-1})$ for convenience. If $m < n$ then the above process is repeated with 1 replaced by some integer not in $\{1, 1\rho,\ldots, 1\rho^{m-1}\}$. In this way ρ is associated with a number of cycles, as permutations of the form (2) are called; and it is easy to see that these cycles involve distinct symbols and that each pair of them commute as permutations. The product of the cycles of ρ is in fact ρ. For example, let ρ be

$$\begin{pmatrix} 1 & 2 & 3 & 4 & 5 & 6 \\ 4 & 2 & 1 & 3 & 6 & 5 \end{pmatrix}.$$

Then we write ρ in the form $(1,4,3)(2)(5,6)$. Often cycles like (2) involving only one symbol are omitted; thus $\rho = (1,4,3)(5,6)$. The decomposition just explained of a permutation into the product of cycles that contain no symbol in common is unique, apart from the order in which the cycles appear.

There is another important way of decomposing permutations. For simplicity consider the special permutation $\rho_0 = (1, 2,\ldots, m)$. It will be found that

$$\rho_0 = (1,2)(1,3)\ldots(1,m).$$

To verify this statement is an easy exercise in multiplying permutations. Permutations of the form (a, b), where $a \neq b$, are called transpositions. Every cycle is the product of suitable transpositions, as we have just seen; but this product is not unique, symbols may be common to different transpositions, and the transpositions in it may not commute. By way of proof we note that

$$\rho_0 = (m, 1)(m, 2)\ldots(m, m-1),$$

$$(12)(13) = (123) \neq (132) = (13)(12).$$

It is now clear, however, that any permutation, not only a cycle, is the product of suitable transpositions.

There is a useful algorithm for calculating $\beta^{-1}\alpha\beta$ when α and β are given permutations. We illustrate its use when α is written in the form

$$\begin{pmatrix} 1 & 2 & \ldots & n \\ 1\alpha & 2\alpha & \ldots & n\alpha \end{pmatrix}.$$

The reader will find it easy to prove that $\beta^{-1}\alpha\beta$ is

$$\begin{pmatrix} 1\beta & 2\beta & \ldots & n\beta \\ 1\alpha\beta & 2\alpha\beta & \ldots & n\alpha\beta \end{pmatrix}.$$

In the case when $\alpha = (1, 2, ..., m)$ we have $\beta^{-1}\alpha\beta = (1\beta, 2\beta, ..., m\beta)$, and products of cycles can be treated on an obvious basis.

All the above remarks apply to quite general permutations; we have not essentially assumed that any permutation affects only a finite number of symbols.

Suppose now that we have a permutation ρ which does have the property just mentioned, so that it leaves fixed all symbols except a finite number, which we take to be $\{1, ..., n\}$. Then ρ can be expressed as the product of suitable transpositions. We define the parity of ρ to be 1 if it is the product of an even number of transpositions, and -1 if it is the product of an odd number; the identity permutation is assigned parity 1. It is clear that if parity is well defined then it has a multiplicative property—namely, if ρ_1, ρ_2 have parities p_1, p_2 respectively then $\rho_1\rho_2$ has parity $p_1 p_2$.

To prove that parity is well defined is a notoriously tricky business.† What has to be established is essentially this: no permutation can be expressed both as the product of an even number of transpositions and as the product of an odd number.

We first investigate the outcome of multiplying a permutation ρ by a transposition τ, with ρ expressed as the product of cycles each containing more than one symbol and no two cycles containing a common symbol. Let $\tau = (a, b)$ with $a \neq b$. There are three cases.

(i) Neither a nor b appears in the expression for ρ. Then $\rho\tau$ has an obvious expression like that for ρ, but with the additional cycle (a, b).

(ii) Just one of $\{a, b\}$, say a, occurs in the expression for ρ. We may write the cycle ρ_0 in which a occurs as $(a, b_1, ..., b_m)$. Then

$$\rho_0 \tau = (a, b_1, ..., b_m)(a, b) = (a, b_1, ..., b_m, b).$$

Thus the effect of right-multiplying ρ by τ is to place an extra symbol in one cycle and to leave the others as they were.

(iii) Both a and b occur in the expression for ρ. If a and b occur in the same cycle multiplication goes like this:

$$\begin{aligned} \rho_0 \tau &= (a, b_1, ..., b_m, b, c_1, ..., c_k)(a, b) \\ &= (a, b_1, ..., b_m)(b, c_1, ..., c_k). \end{aligned}$$

However, a and b may also occur in different cycles:

$$\begin{aligned} \rho_0 \tau &= (a, b_1, ..., b_m)(b, c_1, ..., c_k)(a, b) \\ &= (a, b_1, ..., b_m, b, c_1, ..., c_k). \end{aligned}$$

Thus the number of cycles increases or decreases by one, and the number of symbols appearing is unchanged.

† For a fallacious demonstration on which a proof of invariance of parity is based, see F. M. Hall, *An introduction to abstract algebra*, vol. i, Theorem 12.6.2 (Cambridge University Press, 1966).

This suggests defining a function $P(\rho)$ as follows. Take the standard expression for ρ, involving say n cycles containing $m_1,\ldots, m_n$ symbols respectively, with each $m_i \geqslant 2$. Define

$$P(\rho) = (-1)^{m_1-1}(-1)^{m_2-1}\ldots(-1)^{m_n-1},$$

with $P(\iota) = 1$. Our remarks above show that if τ is any transposition then $P(\rho\tau) = -P(\rho)$.

The way ahead is now clear. Suppose that a permutation ρ can be expressed as the product of r transpositions. An obvious induction based on the last paragraph shows that $P(\rho) = (-1)^r$. But $P(\rho)$ is independent of the expression of ρ in terms of transpositions. It follows that $P(\rho) = 1$ for one such expression if and only if $P(\rho) = 1$ for all such expressions. In other words, ρ cannot have expressions as the product of an even number of transpositions and of an odd number.

It follows that parity is well defined. In fact the function $P(\rho)$ described above is the parity of ρ. A permutation is said to be even or odd according as its parity is 1 or -1.

Consider next the set R of all permutations that move only a finite number of symbols in some set. Let R_e, R_o denote the subset of even, odd permutations respectively. We wish to prove that these subsets are in one–one correspondence, assuming that the trivial case when R_o is empty is avoided. Let τ be any transposition in R_o. Then any element of R_o may be expressed as $\rho\tau$, for some $\rho \in R_e$; the desired correspondence is that in which ρ and $\rho\tau$ are matched. Note that in particular there are $\frac{1}{2}n!$ even and $\frac{1}{2}n!$ odd permutations on a finite set with n elements, provided $n > 1$.

The concept of a mapping of a set into itself admits a fruitful generalization in the concept of a relation on a set. We shall define only the notation, not the concept. If x, y are elements of a set S and x is related to y then we write $x \sim y$. We note that the conditions $x \sim y$ and $x \sim z$, for some $z \in S$, do not together imply $y = z$, as in the case of a mapping; but that a relation does have this in common with a mapping, that it determines a subset of $S \times S$.

Equivalence relations are of special interest in mathematics. The basic reason for this is that it is often much easier to say when two objects have the same property than to describe the property; consider the property of being a right-handed (not left-handed) set of axes, or of having a certain number of elements. Equivalence relations are the tool used in comparing objects in respect of their properties.

To be specific, the relation $\sim$ is an equivalence relation on S if and only if it has the following three features.

(i) The reflexive property: $x \sim x$.

(ii) The symmetric property: if $x \sim y$ then $y \sim x$.

(iii) The transitive property: if $x \sim y$ and $y \sim z$ then $x \sim z$.

Here x, y, z are arbitrary elements of S. Note that (i) may be written as an implication, like (ii) and (iii), for it is equivalent to: if $x = y$ then $x \sim y$.

A most important equivalence relation is that of congruence between integers. The set concerned is the set Z of integers. Let n be a fixed positive integer. We define the relation $x \equiv y$, where $x, y \in Z$, to mean that $x-y$ is divisible by n. The reader will find that verification of (i), (ii), and (iii) requires only the most elementary properties of integers.

Suppose we have a partition of the set S. It should be clear that we obtain an equivalence relation on S by defining $x \sim y$ to mean that x and y belong to the same element of the partition. The basic theorem about equivalence relations is a converse of this: every equivalence relation on S determines a partition of S.

To prove this we introduce the so-called equivalence classes. We define $[a]$ to be $\{x : x \sim a\}$ for each a in S. Now each element of S lies in some class, because $a \in [a]$ by (i). We therefore require to prove that no element of S lies in two distinct classes. Suppose then that $a \in [b]$ and $a \in [c]$. By definition of equivalence class we have $a \sim b$ and $a \sim c$. By (ii) we have $b \sim a$ and $a \sim c$. By (iii) we have $b \sim c$. Now if $x \in [b]$ then $x \sim b$, and so $x \sim c$ by (iii), that is $x \in [c]$. Therefore $[b] \subseteq [c]$, similarly $[c] \subseteq [b]$, and finally $[b] = [c]$. We have shown this: either $[b] \cap [c] = \varnothing$ or $[b] = [c]$. The theorem stated in the last paragraph is therefore established.

Consider the equivalence classes defined in Z by congruence modulo n. We see that

$$[a] = \{x : x \equiv a \text{ modulo } n\}.$$

Therefore $[a]$ may be described as the set

$$\{..., a-n, a, a+n, a+2n, ...\}.$$

The distinct equivalence classes are $[0], ..., [n-1]$; note that $[n] = [0]$. We often refer to these classes as the residue classes modulo n, and we call $[0]$ the zero residue class. (A factor group is a set of equivalence classes with a certain structure, and more generally a coset is an equivalence class.)

We mention a relation of quite a different sort from an equivalence relation. A total ordering on a set S will be written thus: $x \leqslant y$ if x, y are elements of S which the total ordering relates. We assume that if $x, y \in S$ then either $x \leqslant y$ or $y \leqslant x$ must hold, in addition to:

(i) $x \leqslant x$;

(ii) if $x \leqslant y$ and $y \leqslant x$ then $x = y$;

(iii) if $x \leqslant y$ and $y \leqslant z$ then $x \leqslant z$.

The natural ordering of the real numbers leads to the most familiar examples of total orderings.

Our final consideration in this Appendix concerns the beginnings of number theory. We take it for granted that every integer greater than 1 can be expressed as the product of primes, uniquely apart from the order of the primes. It thus makes sense to talk of the greatest common divisor of two positive integers; if we suppose it to be positive then it is uniquely defined. Two such integers are said to be coprime (or relatively prime) when their greatest common divisor is 1. We shall also assume the division algorithm: if d is a positive integer and c is any integer then there are integers γ, δ for which $c = \gamma d + \delta$ and $0 \leqslant \delta < d$.

The so-called Euclidean algorithm is often referred to in elementary group theory. Its basic form asserts that if the non-zero integers a, b are coprime then there are integers α, β (not necessarily positive) for which

$$\alpha a + \beta b = 1.$$

We prove this. Our proof consists in a deduction of the Euclidean algorithm from the statement that every non-empty set of positive integers contains a least integer (which also implies the division algorithm, by the way). Consider the set S of numbers,

$$S = \{\lambda a + \mu b : \lambda \in Z, \mu \in Z\}.$$

Clearly S contains a non-empty subset of positive integers, and we suppose that d is the least of these. Let α, β be elements of Z such that $\alpha a + \beta b = d$. Suppose further that c is any element of S, then $\lambda_0 a + \mu_0 b = c$ for some λ_0, μ_0 in Z. By the division algorithm there are integers γ, δ such that $c = \gamma d + \delta$ and $0 \leqslant \delta < d$. Now $\delta \in S$ as the following calculation shows:

$$\begin{aligned} \delta &= c - \gamma d \\ &= (\lambda_0 a + \mu_0 b) - \gamma(\alpha a + \beta b) \\ &= (\lambda_0 - \gamma\alpha)a + (\mu_0 - \gamma\beta)b. \end{aligned}$$

The facts that d is the minimal positive element in S and that $0 \leqslant \delta < d$ now show that $\delta = 0$. That is, d divides c, and so the special element d divides every element of S. It is clear that a and b are in S (take $\lambda = 1$, $\mu = 0$ and $\lambda = 0$, $\mu = 1$ respectively). Therefore d divides the two coprime integers a and b, and so $d = 1$. This completes the desired proof.

A generalization is at once possible: if h is the greatest common divisor of the non-zero integers a and b, then there are integers α, β such that

$$\alpha a + \beta b = h.$$

The idea of the proof of this is to apply the preceding result to the integers a/h, b/h; the reader will be able to fill in the details.

A further generalization concerns the case of n non-zero integers $a_1, \ldots, a_n$ whose greatest common divisor is 1. It is then possible to find integers $\alpha_1, \ldots, \alpha_n$ such that

$$\alpha_1 a_1 + \ldots + \alpha_n a_n = 1.$$

The proof is by induction on n and uses the now known result for $n = 2$. Suppose that $n > 2$ and that the result has been established for $n-1$. If the greatest common divisor of $a_1, \ldots, a_{n-1}$ is h then h and a_n are coprime, in accordance with the hypothesis. Further, the greatest common divisor of $a_1/h,\ldots, a_{n-1}/h$ is 1, and the inductive hypothesis gives integers $\alpha'_1,\ldots, \alpha'_{n-1}$ such that

$$\alpha'_1 a_1+\ldots+\alpha'_{n-1} a_{n-1} = h.$$

It also gives integers β_1, β_2 such that

$$\beta_1 h+\beta_2 a_n = 1.$$

We take $\alpha_1 = \alpha'_1 \beta_1,\ldots, \alpha_{n-1} = \alpha'_{n-1}\beta_1, \alpha_n = \beta_2$, and the inductive step is complete.

Little more than the definition is needed as regards the use made of complex numbers in this book. We remark that any such number may be expressed in the form $a+ib$ where a, b are real and i is a particular complex number for which $i^2 = -1$. Alternatively, a non-zero complex number may be expressed as $r(\cos\theta+i\sin\theta)$ where $r > 0$ and $0 \leqslant \theta < 2\pi$, r being called its modulus. We note de Moivre's theorem:

$$\{r(\cos\theta+i\sin\theta)\}^n = r^n(\cos n\theta+i\sin n\theta),$$

for any integer n. There is an easy proof, using induction on n combined with a special argument for the case $n = -1$.

Note on Further Reading

THE present book on group theory is essentially introductory. After mastering it the young mathematician should be ready for the more specialized works listed below; however, of the more advanced *general* works still in circulation we mention the books of Kurosh and Hall, emphasizing infinite and finite group theory respectively.

The student of finite groups will have to learn a great deal about group representations, on which the book of Curtis and Reiner is an excellent treatise. Huppert is writing a very comprehensive work on finite groups, including lengthy accounts of representations, cohomology, permutation groups, etc.; the first volume on finite soluble groups has just become available, and a second volume on finite simple groups is promised.

A thorough grounding in ordinals and cardinals, as well as use of equivalents of Zorn's lemma, is necessary before more about infinite groups can be tackled; this material may be learnt from the many good available books on infinite set theory. The book of Magnus, Karrass, and Solitar will then be found first-rate for topics which are generalizations of freeness in groups and of properties defined by series. More special, but none the less an attractively written book, is Hanna Neumann's monograph on varieties of groups.

Wielandt's work affords an introduction to permutation groups, Fuchs' book is a standard work on abelian groups, and that of Dickson, though old, is still a good account of linear groups.

No heed has been taken in the present book of applications of groups to physics, and we may remark that it is *representations of topological groups* that are usually required; good references here are the books of Boerner and Hamermesh, the former being on the mathematical side and the latter more physical.

H. BOERNER, *Representations of groups with special consideration for the needs of modern physics* (North-Holland, 1963).

C. W. CURTIS and I. REINER, *Representation theory of finite groups and associative algebras* (Interscience, 1962).

L. E. DICKSON, *Linear groups with an exposition of the Galois field theory* (Dover, 1958).

L. FUCHS, *Abelian groups* (Publishing house of the Hungarian Academy of Sciences, 1958).

M. HALL, *The theory of groups* (Macmillan, 1959).

M. HAMERMESH, *Group theory and its application to physical problems* (Addison-Wesley, 1962).

B. HUPPERT, *Endliche Gruppen* I (Springer-Verlag, 1967).

A. G. KUROSH, *The theory of groups*, vols. i and ii (Chelsea, 1955, 1956).

W. MAGNUS, A. KARRASS, and D. SOLITAR, *Combinatorial group theory* (Interscience, 1966).

H. NEUMANN, *Varieties of groups* (Springer-Verlag, 1967).

H. WIELANDT, *Finite permutation groups* (Academic Press, 1964).

Since the above was written a number of new texts have appeared, and the most significant in the present context are listed below. The work of Gorenstein is on a scale comparable to Huppert's book. Passman gives a more extensive treatment of permutation groups than Wielandt, while the works concerned with representation theory could be arranged in ascending order of difficulty as follows: Herstein, Burrow, Feit.

M. BURROW, *Representation theory of finite groups* (Academic Press, 1965).

W. FEIT, *Characters of finite groups* (Benjamin, 1967).

D. GORENSTEIN, *Finite groups* (Harper and Row, 1968).

I. HERSTEIN, *Noncommutative rings* (The Mathematical Association of America, distributed by Wiley, 1968).

D. PASSMAN, *Permutation groups* (Benjamin, 1968).

Index